ELEMENTARY CLASSICAL PHYSICS II
Laboratory Manual

Jutta Luettmer-Strathmann
The University of Akron Department of Physics

VAN-GRINER

Elementary Classical Physics II

Laboratory Manual
Jutta Luettmer-Strathmann
The University of Akron Department of Physics

I would like to thank the students and teaching assistants whose comments and suggestions helped shape the Physics 292 laboratory. Special thanks go to Dr. Ivan Dolog, who shared his expertise in experimental physics and photography.

Printed in the United States of America
10 9 8 7 6 5 4 3
ISBN: 978-1-61740-493-1

Van-Griner Publishing
Cincinnati, Ohio
www.van-griner.com

CEO: Mike Griner
President: Dreis Van Landuyt
Project Manager: Brenda Schwieterman
Customer Care Lead: Julie Reichert

Strathmann 493-1 Su18
183772-321902
Copyright © 2019

PHYSICS 292

TABLE OF CONTENTS

INTRODUCTION

OBJECTIVES

The purpose of this physics laboratory is to provide an experimental foundation for theoretical concepts discussed in class and to introduce you to important physical phenomena and principles not (yet) treated in the lectures. During this laboratory you will

- learn experimental methods to investigate physical phenomena,
- develop skills in acquiring and presenting experimental data,
- estimate uncertainties in measured and derived quantities (error propagation),
- interpret experimental data within a theoretical framework and test theoretical predictions, and
- write laboratory reports that describe experimental design, observations, data analysis, and conclusions in a clear and concise manner.

BEFORE THE LAB

Please prepare for the experiments by reading the description of the experiment as well as the appropriate section(s) in your textbook. In general, you are expected to complete a PreLab quiz, please see the Syllabus for details.

Please bring the following items to each lab session:

- Lab manual
- Clean filler paper, pencil, and pen
- Flash drive (USB memory stick)

During the Lab

Each group will collect their observations, data, graphs, and answers to questions as requested by the lab manual. Please be sure to write the date and the initials of the group members at the top of each page you use. In order to make the laboratory session a good learning experience for you and your colleagues, be prepared to

- participate actively in the lab,
- be on time for the lab,
- not have food or drink in the lab,
- turn off your cell phone,
- be considerate of the other students in the lab,
- be cautious with lab equipment,
- *leave your lab area neat and clean,* and
- copy your Excel and LinReg files to your flash drive and *remove them from the hard drive* before you leave.

These criteria are used to determine your grade for Performance in the Lab.

At the End of the Lab Period

Submit your answers to questions in the lab description; they will be graded for a group grade. Have the data pages and printouts initialized by the instructor; these pages become part of the laboratory report and should be taken home by the student writing the report for the experiment. When no lab report needs to be written, submit these pages together with your lab-question answers.

After the Lab

The teaching assistant will assign three laboratory reports to each student to write for an individual grade.

Please Note

- Each student must pass the lab part of the course in order to pass the course.
- A student who misses three or more lab sessions fails the lab.
- No student can make up more than one lab in a make-up week.
- Make-ups can only be arranged with the instructor's approval.
- If you know ahead of time that you will have to miss a lab, notify your instructor as soon as possible and discuss a make-up.

LABORATORY REPORTS

It may be helpful to think about an audience when you write your reports. Imagine your company considers buying a new method or piece of equipment and you are the engineer in charge of testing. Your report would be read by very busy people who would hold you responsible if something went wrong later; this means the report has to be ***complete and concise.***

Please prepare your lab reports with a word processor. The following table shows the sections that are standard for the reports (points for each section are noted in parentheses).

1. (1)	TITLE SECTION	Date: Lab section: Name of TA: Your name Name of your partners Title of the experiment
2. (7)	INTRODUCTION AND THEORY	State the purpose of the experiment and introduce the pertinent physical concepts to help prepare the reader for what follows. Define the variables representing physical quantities and explain the equations used in the evaluation of the experiment.
3. (5)	PROCEDURE	Give a brief explanation how the data were taken. Sketches of the set up may be essential to get your idea across. Another person should be able to reproduce your experimental data based on your account.
4. (10)	DATA AND GRAPHS	Supply **all** data! Anything you measure or anything the computer measures and calculates for you is considered data. Please be sure to include all of the original data sheets and printouts in your report. Numerical values should be in the form of tables and, when appropriate, graphs. All graphs should have a number, a title, and clearly labeled axes with appropriate units. If you derive information from a graph (slope, maximum value, etc.) indicate it on the graph as well as in the "RESULTS" section.
5. (10)	CALCULATIONS AND RESULTS	All equations used in calculations have to be introduced in the "THEORY" section. Calculations should be clear and complete, and results should be boxed to stand out. If you use Excel to perform calculations, copy the appropriate columns into your lab report. Please be sure that the column titles contain the necessary information (name of variable (with units), equation for calculations) and are clearly visible. State the results obtained from computer printouts referring to the graphs by number.
6. (7)	CONCLUSIONS	Restate the purpose of the lab and summarize your results. Explain if the results make sense and discuss why or why not. Discuss sources of experimental uncertainty and suggest improvements if you can.

Error Estimates

Precision and accuracy refer to two different types of experimental uncertainty—random (or statistical) and systematic errors. Often, you can estimate the uncertainty due to random errors by performing an experiment many times and evaluating the data with statistical tools. Sometimes you can estimate this uncertainty from the type of equipment you are using. Detecting systematic errors is not straightforward; if you suspect systematic errors in your experiments, discuss them in the lab report.

Random Errors

For repeated measurements y_k of a quantity y, the random error may be estimated from the standard deviation:

$$\sigma^2 = \frac{1}{N-1} \sum_{k=1}^{N} (y_k - y)^2$$

$$\sigma = +\sqrt{\sigma^2}$$

Standard Deviation (1)

Here y denotes the average value. In Excel, the function STDEV calculates the standard deviation. If repeated measurements give the same value, the random error is **not zero,** but given by the resolution of the equipment (for example, a typical ruler has a resolution of 0.5–1 mm).

Error Propagation

We often perform measurements to determine quantities that are then used to calculate other quantities. If a calculated quantity h depends on the measured quantities $r, s, t, \dots$ each with their own uncertainties $\sigma_r, \sigma_s, \sigma_t, \dots$ then the uncertainty σ_h of $h(r, s, t, \dots)$ is calculated from the partial derivatives according to

$$\sigma_h^2 = \left(\frac{\partial h}{\partial r}\right)^2 \sigma_r^2 + \left(\frac{\partial h}{\partial s}\right)^2 \sigma_s^2 + \left(\frac{\partial h}{\partial t}\right)^2 \sigma_t^2 + \dots$$

$$\sigma_h = +\sqrt{\sigma_h^2}.$$

Propagated Uncertainty (2)

In Equation (2), the expression $\left(\dfrac{\partial h}{\partial r}\right)$ stands for the partial derivative of the function h with respect to the variable r. This means that all other variables ($s, t, \dots$) are treated as constants when the derivative is calculated.

Two examples for error propagation are shown on the next pages.

ABSOLUTE AND RELATIVE UNCERTAINTY

The uncertainty σ is also called the ***absolute uncertainty;*** it has the same units as the result of the experiment. By itself, it may not mean so much (a \$10.00 uncertainty in the price of a house is probably acceptable, whereas the same uncertainty in the price of a movie ticket probably is not). The **relative uncertainty** of an average value y, typically measured in percent, is often more helpful.

$$\frac{\sigma_y}{y} = \frac{\sigma_y}{y}\,100\% \qquad\qquad\qquad \textbf{\textit{Relative Uncertainty}} \quad (3)$$

EXAMPLE 1

The measured frequency is $f = 1000$ Hz with uncertainty $\sigma_f = 0.5$ Hz. What is the uncertainty σ_T in the period $T = 1/f$?

We have $T = 1/f = 10^{-3}$ s $= 1.0$ ms; to find the uncertainty in T, substitute T for h and f for r in Equation (2). (There are no variables s and t in this case.)

$$\sigma_T^2 = \left(\frac{\partial T}{\partial f}\right)^2 \sigma_f^2 = \left(\frac{\partial(1/f)}{\partial f}\right)^2 \sigma_f^2 = \left(-\frac{1}{f^2}\right)^2 \sigma_f^2 \Rightarrow$$

$$\sigma_T = \frac{1}{f^2}\,\sigma_f = 10^{-6}\,\text{s}^2 \times 0.5\,\text{s}^{-1} = 5 \times 10^{-7}\,\text{s}$$

The result for the period is $T = (1.0 \pm 5 \times 10^{-4})$ ms.

It is instructive to calculate the relative uncertainties.

$$\frac{\sigma_f}{f} = \frac{0.5\,\text{Hz}}{1000\,\text{Hz}}\,100\% = 0.05\% \quad\text{and}\quad \frac{\sigma_T}{T} = \frac{5\cdot 10^{-4}\,\text{ms}}{1.0\,\text{ms}}\,100\% = 0.05\%$$

The relative uncertainties are the same in this case, since the period is the ***first*** inverse power of the frequency. You should convince yourself that the relative uncertainty of y^2 (or y^{-2}) is twice the relative uncertainty of y, and that the relative uncertainty of the nth power is n times the relative uncertainty of the first power of a quantity.

EXAMPLE 2

The time constant τ of an RC circuit is related to the resistance R and the capacitance C through $\tau = RC$. You measured $\tau = (0.034 \pm 0.0038)$ s and $R = (100 \pm 0.5)\ \Omega$, which gives for the capacitance $C = \tau/R = 3.4 \times 10^{-4}\ \text{s}/\Omega = 3.4 \times 10^{-4}$ F. What is the uncertainty in the capacitance?

$$\sigma_C^2 = \left(\frac{\partial C}{\partial \tau}\right)^2 \sigma_\tau^2 + \left(\frac{\partial C}{\partial R}\right)^2 \sigma_R^2 = \left(\frac{1}{R}\right)^2 \sigma_\tau^2 + \left(-\frac{\tau}{R^2}\right)^2 \sigma_R^2$$

$$= \left(\frac{1}{100\ \Omega}\right)^2 (0.0038\ \text{s})^2 + \left(\frac{0.034\ \text{s}}{10{,}000\ \Omega^2}\right)^2 (0.5\ \Omega)^2$$

$$= 1.4 \cdot 10^{-9}\,\frac{\text{s}^2}{\Omega^2} \Rightarrow \sigma_C = 3.8 \cdot 10^{-5}\ \text{F}$$

The result for the capacitance is $C = (340 \pm 38)\ \mu\text{F}$. Note that the uncertainty is dominated by the error in τ, as the relative uncertainties show:

$$\frac{\sigma_\tau}{\tau} = \frac{0.0038\ \text{s}}{0.034\ \text{s}}\,100\% = 11\%, \quad \frac{\sigma_R}{R} = \frac{0.5\ \Omega}{100\ \Omega}\,100\% = .05\%, \quad \text{and}$$

$$\frac{\sigma_C}{C} = \frac{38\ \mu\text{F}}{340\mu\text{F}}\,100\% = 11\%$$

USEFUL EQUATIONS

The following are useful equations for propagating errors (r and s are measured quantities, C and a are constants):

$$\sigma_{r+s} = \sqrt{\sigma_r^2 + \sigma_s^2}, \quad \sigma_{r-s} = \sqrt{\sigma_r^2 + \sigma_s^2}, \quad \sigma_{r\times s} = \sqrt{s^2\sigma_r^2 + r^2\sigma_s^2},$$

$$\sigma_{r/s} = \sqrt{\frac{1}{s^2}\sigma_r^2 + \frac{r^2}{s^4}\sigma_s^2}, \quad \sigma_{Cr} = C\sigma_r, \quad \sigma_{Cr^a} = Car^{a-1}\sigma_r,$$

$$\sigma_{C\exp(ar)} = Ca\exp(ar)\sigma_r, \quad \sigma_{C\ln(ar)} = \frac{C}{|r|}\sigma_r.$$

LINREG REFERENCE

LinReg is software that allows you to plot data and perform linear fits that take the uncertainty of the input data into account.

x	sigma_x	y	sigma_y
15	0.15	9.7	0.49
21	0.21	4.4	0.22
25	0.25	3.2	0.16

PREPARING EXCEL DATA FOR USE IN LINREG

Prepare a clean Excel worksheet with data (x and y) and uncertainties (sigma_x and sigma_y) arranged in four columns, as in the table above. Save the worksheet as a "Text (Tab delimited) *.txt" file with a new name, for example neatdata.txt. Close the Excel file with the name neatdata.txt before you proceed.

ENTERING DATA IN LINREG

Start the program LinReg by double clicking on the icon and enter information on the opening page (your name(s), name and units of the quantities that will be plotted on the x- and y-axes). Click on the "Data" tab, go to the "File" menu and choose "Import." Navigate to your text file with the data (e.g., neatdata.txt) and import the data as four columns separated by tabs. Change the Table format to see more digits, if necessary, and clear the zeros in the first row. If you have a small amount of data, you can also type the data directly into the "Data" columns.

PLOTTING AND FITTING DATA WITH LINREG

Click on the "Graph" tab and choose "Do Linear Fit"; the slope and intercept will be displayed along with error estimates based on the analysis of simulated data within your error bars (see "Help" for details). Adjust the axis limits, if necessary, and click on "Edit Graph Labels" to modify the title of the graph before printing your results. Before entering new data, remember to save your file by choosing "Save As" from the "File" menu.

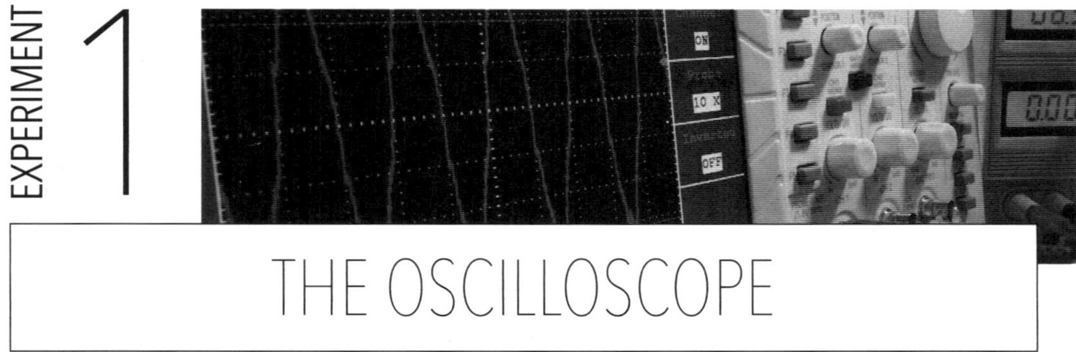

THE OSCILLOSCOPE

OBJECTIVE

◎ To learn how to use an oscilloscope to display a periodic signal and measure its frequency.

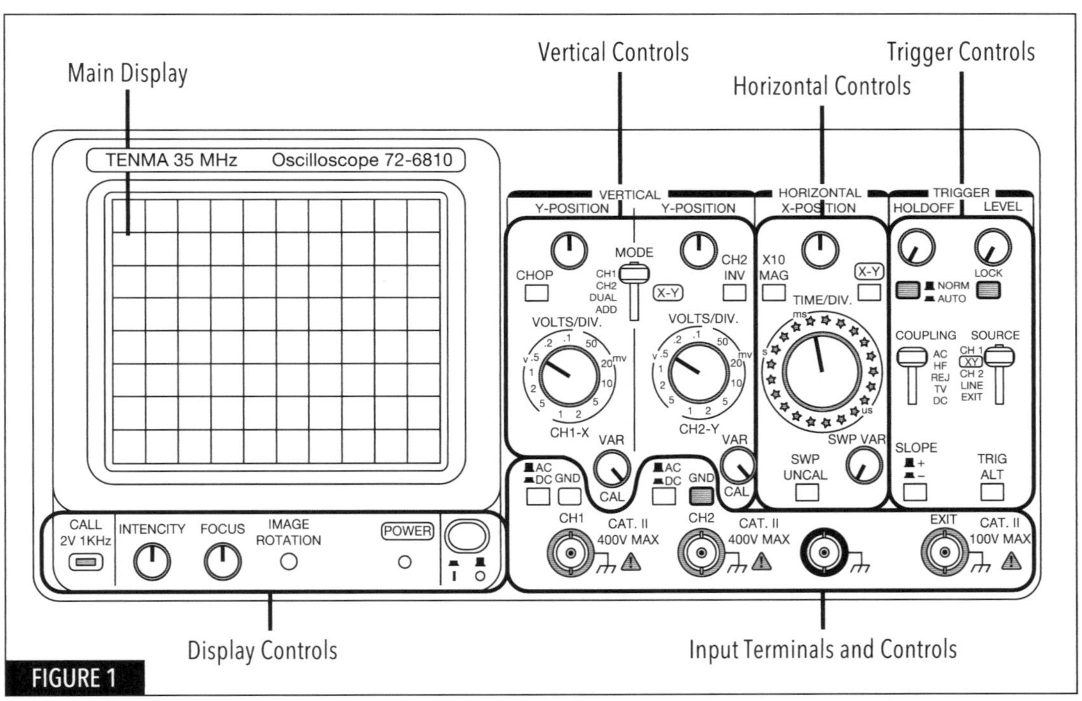

Front of the oscilloscope with the controls in their initial settings. A dark rectangle indicates a switch that is pressed in, positions of knobs and sliding switches are indicated.

INTRODUCTION

An oscilloscope is a device to display electrical signals on a screen. A cathode ray oscilloscope works much like an old-fashioned, black-and-white television set: An electron beam is generated in the back of a vacuum tube and focused onto a fluorescent screen at the front, where it creates a bright spot. Two sets of deflecting plates along the tube allow the beam to be moved horizontally and vertically. We will use the oscilloscope in a mode where the beam sweeps automatically from left to right at a constant velocity (and then jumps back again) so that the *x*-axis of the display becomes a time axis. The oscilloscope has five main types of controls (Display, Vertical, Horizontal, Trigger, and Input) as shown in Figure 1. We will explore the controls that are important for the experiments in this laboratory.

LAB EQUIPMENT

◎ Oscilloscope

◎ Digital function generator

PROCEDURE

PART I: OSCILLOSCOPE ONLY

1. Set the knobs and switches to the initial settings indicated in Figure 1, turn the oscilloscope on, and wait for 30 seconds. Adjust the focus and intensity (Display Controls) until you have an easily visible trace on the screen and use the position knobs (Vertical Display, Channel 1, and Horizontal Display) to center the trace.

2. Turn the TIME/DIV knob (Horizontal Controls) all the way to the left (0.5 s) and notice how slowly the spot on the screen is moving. With the SWP.UNCAL button released, the setting of the TIME/DIV knob shows how much time it takes the beam to cross one division (one square = 1 cm) in the horizontal direction. Since there are ten squares, it should take the beam 10×0.5 s = 5.0 s to cross the screen. If the SWP.UNCAL button is pressed, the SWP.VAR knob can be used to change the horizontal scale (try it, but don't forget to reset the switch). This is useful for some experiments but it means the horizontal calibration is lost. Keep the SWP.UNCAL button released for today's experiments.

3. Now turn the TIME/DIV knob step by step to the right until you reach 0.5 ms. You should see the spot on the screen speed up and become a straight line since our eyes cannot resolve the fast motion of the spot.

Question 1   What is the speed of the spot on the screen when the TIME/DIV knob is set to 0.5 s? Express your answer in divisions/second and in meter/second.

Part II: Periodic Signal

1. Connect the output of the function generator to the banana-plug adaptor on the Channel 1 input socket (Input Controls) making sure you connect red-to-red and black-to-black.

2. Turn the voltage knob all the way to the left (zero amplitude), turn the function generator on, and set the function generator to produce a harmonic (sine/cosine) signal of frequency 500 Hz. Now gradually increase the amplitude until you see a clear trace of a sine/cosine function on the screen.

3. With the VAR knob (Vertical Controls, Channel 1) in the CAL position (calibrated), the setting of the VOLTS/DIV knob shows how many volts correspond to one vertical division. Adjust the vertical scale (VOLTS/DIV knob) until the amplitude of your signal is about three divisions, then adjust the amplitude on the function generator to make it exactly three divisions.

4. Adjust the TIME/DIV knob (Horizontal Controls) until you see about two periods of the signal on the screen. Adjust the horizontal position of the signal (Horizontal Controls) to make the signal easy to evaluate.

5. Sketch the displayed pattern on a piece of paper, indicate the amplitude and the period in your sketch, and note the amplitude A (in divisions and in Volt). Record the period T of the signal (in divisions and in seconds) and estimate the uncertainty σ_T of the period. Make a table in Excel for your data as shown below (you may write sigma_A for σ_A etc. in Excel).

Set Frequency: 500 Hz					
A (div)	sigma_A (div)	scale (V/div)	A (V)	sigma_A (V)	
T (div)	sigma_T (div)	scale (s/div)	T (s)	sigma_T (s)	sigma_T/T

Question 2 The frequency f of a signal is related to the period T by $f = 1/T$. Find an equation for the uncertainty of the frequency σ_f by error propagation from the uncertainty σ_T in the period.

6. In Excel, calculate the frequency $f = 1/T$ and the uncertainty of the frequency σ_f, adding columns as shown below. Repeat Steps 3–6 for a frequency of 1000 Hz.

f (Hz)	sigma_f (Hz)	sigma_f/f

Question 3 Do your values for the uncertainties satisfy $\sigma_f/f = \sigma_T/T$? If not, find the mistake and correct the calculation.

Question 4 To compare your calculated frequency with the frequency displayed by the function generator, calculate the difference between the values and compare the difference with the calculated uncertainty σ_f. Do the displayed and calculated frequencies agree within the uncertainty?

Question 5 Are there other uncertainties that should be included?

Please submit the answers to the questions, your sketches, and a printout of the calculations at the end of the lab period.

VIBRATING STRINGS

OBJECTIVES

- To measure how the resonant frequencies of a stretched string depend on the applied tension.
- To determine a string's linear mass density.

INTRODUCTION

Stretched strings form the basis of many musical instruments such as the guitar, the violin, and the piano. A short time after a string has been plucked its vibrations are dominated by the lowest resonant frequency of the string. If you play the guitar, you know that you can change this frequency by changing the tension in the string (tuning), by changing the vibrating length (fingering), and by replacing the string with one of a different mass per unit length. The frequency is called a resonant frequency because the string will vibrate with maximum amplitude (resonate) when it is excited by a signal with this frequency.

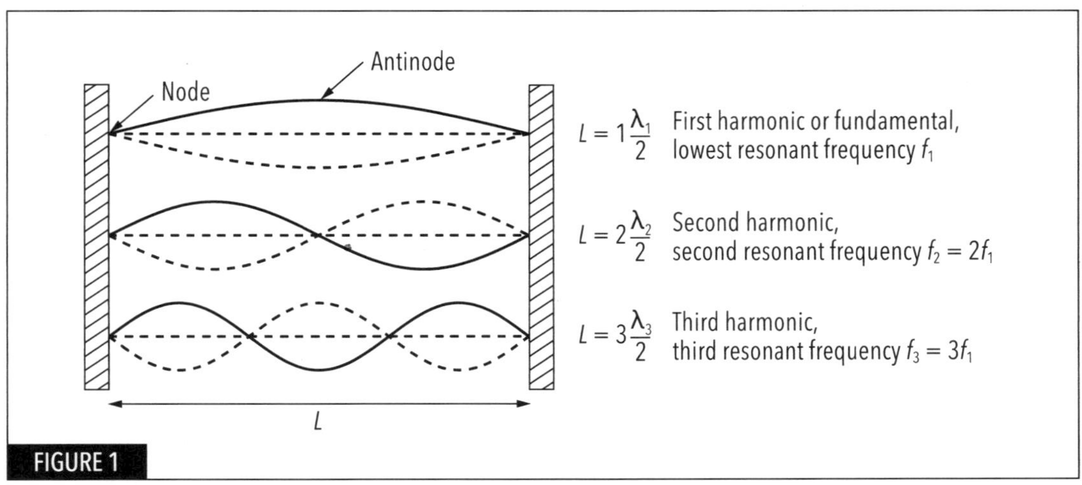

FIGURE 1

Standing waves corresponding to the lowest three resonant frequencies of a string.

The resonant vibrations of a string correspond to standing waves. A standing wave is a superposition of two waves of the same frequency and amplitude that travel in opposite directions. For a resonant vibration on a string that is fixed at both ends, the wave pattern has to have nodes at both ends. This is the case when multiples of half a wavelength fit on a string as shown in Figure 1. The largest wavelength is $\lambda_1 = 2L$, where L is the vibrating length of the string, the second largest is $\lambda_2 = L$, and in general we write

$$\lambda_n = \frac{2L}{n}, \quad n = 1, 2, ..., \tag{1}$$

where n is called the order of the harmonic.

The wavelength λ is related to the frequency f and the speed v of a wave through

$$v = \lambda f. \tag{2}$$

Since the wave speed is the same for all harmonics we have $v = \lambda_1 f_1 = \lambda_2 f_2 = ... = \lambda_n f_n$ and

$$f_n = \frac{v}{\lambda_n} = \frac{nv}{2L}. \tag{3}$$

For waves on a string, the speed of a wave depends only on the tension F_T in the string and the linear mass density μ (the mass of the string divided by its length):

$$v = \sqrt{\frac{F_T}{\mu}}. \tag{4}$$

Question 1 Derive the following equation:

$$f_n = nf_1. \tag{5}$$

What kind of graph do you expect if you plot the resonant frequencies of a string as a function of the order of the harmonic?

Question 2 Derive

$$f_1 = \frac{1}{2L}\sqrt{\frac{F_T}{\mu}}. \tag{6}$$

How can you determine the mass density μ from a graph of f_1^2 as a function of F_T?

LAB EQUIPMENT

- Sonometer with driver and detector coils, bridges, and tensioning lever with five notches
- Digital function generator
- Oscilloscope
- Steel strings
- Weights

Note: *The actual driving frequency is twice the frequency displayed by the function generator.*

PROCEDURE

The sonometer consists of a base with a tensioning lever on one side and a tension adjustment screw (not shown) on the other side. A string is fixed to the sonometer and placed across two bridges which determine the vibrating length of the string. An inductive coil (driver coil), placed about 5 cm from a bridge, is attached to a function generator and excites vibrations in the string. A second inductive coil (detector coil), placed about midway between the bridges, measures the

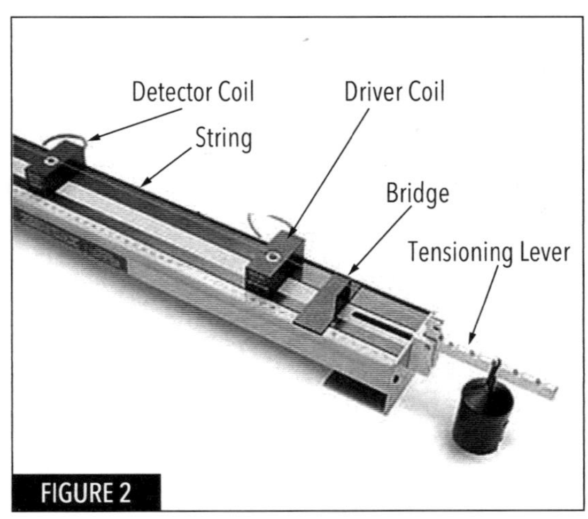

Sonometer.

amplitude of the vibrations and is attached to an oscilloscope which displays the recorded signal. The tensioning lever has five notches which allow the tension in the string to be changed. The relation between the notch N and the tension is

$$F_T = NMg. \tag{7}$$

Here N is the number of the notch, for example, if you hang a mass M from the notch closest to the string, the tension of the string is equal to Mg, where $g = 9.8$ m/s^2 is the acceleration due to gravity. If you hang the mass from notch two, the tension equals $2Mg$, if you hang it from notch three, the tension is $3Mg$, etc. The lever must be horizontal to accurately determine the string tension from the hanging mass.

1. Choose a string from one of the five gauges available (gauges 0.010″, 0.014″, 0.017″, 0.020″, and 0.022″ corresponding to mass densities 0.39 g/m, 0.78 g/m, 1.12 g/m, 1.50 g/m, and 1.84 g/m). Hook one end of the string into the retainer

on the tensioning lever, span the string across the sonometer, and attach the other end to the screw head. Place the bridges under the string so that they are about 60 cm apart. Measure and record this length in an Excel worksheet.

2. Hang a mass (approximately 0.2–0.5 kg) from a notch on the tensioning lever and, if needed, adjust the string adjustment screw to make the tensioning lever horizontal. Record the mass and the notch in your Excel worksheet; calculate the tension, F_T, and estimate an error, σ_{F_T}.

3. Set the function generator to produce a sine wave and set the driving frequency to a value between 20 and 100 Hz; if the string produces a loud sound, reduce the amplitude. Now vary the driving frequency slowly. When you reach a resonant frequency, you should be able to see the motion of the string, the sound produced by the vibrating string should be a maximum, and the wave pattern on the oscilloscope should become a clean sine wave. If you can't see or hear the string, raise the amplitude of the function generator output slightly and try again. The first time you find a resonance, observe the wave pattern to determine the order of the harmonic. Record the order of the harmonic and the frequency in your Excel worksheet (remember the actual frequency is twice the frequency displayed by the generator).

4. Vary the frequency until you have found the first five harmonics.

5. To test if you found all resonances, use Excel to plot the resonant frequencies as a function of the order of the harmonic and make additional measurements if necessary. Add a linear trendline to the Excel graph and record the values of the slope, the intercept, and R^2.

6. Change the tension by moving the mass to another notch (make adjustments to keep the lever horizontal, if necessary), then find the first five harmonics as before. Repeat these steps until you have results for four different tensions.

EVALUATION

The linear mass density will be your main result. You will need several steps to find it and its uncertainty.

1. For each tension F_T, determine the resonant frequency from the slope of the Excel graph and estimate the uncertainty from the R^2 value with the equation:

$$\sigma_{f_1} = f_1 \sqrt{\frac{1 - R^2}{R^2}} \tag{8}$$

Make a clean table in a new worksheet in Excel with the following columns:

$$F_T, \ \sigma_{F_T}, \ f_1, \ \sigma_{f_1}.$$

2. Start the program LinReg, enter your data $(F_T, \sigma_{F_T}, f_1, \sigma_{f_1})$ and create a plot of f_1^2 as a function of F_T. Check that the graph makes sense (if not, look for mistakes), perform a linear fit, choose a good title and adjust axis labels if necessary, then print your results.

3. Calculate the linear mass density μ from the slope of the graph. Estimate the uncertainty σ_L in the vibrating length and calculate the uncertainty σ_μ from the propagated error, see Equation (9).

4. Compare your result for the mass density with the listed values and determine the gauge of your string.

Question 3 Use error propagation to show that

$$\sigma_\mu^2 = \mu^2 \left[\left(2\,\frac{\sigma_L}{L} \right)^2 + \left(\frac{\sigma_s}{s} \right)^2 \right], \tag{9}$$

where s is the slope of the graph of f_1^2 as a function of F_T and σ_s is the uncertainty of the slope.

3

RESONANCE FREQUENCIES OF A TUBE

OBJECTIVES

- ◎ To measure the resonant frequencies of an open tube and determine the speed of sound in air.
- ◎ To measure the lengths for which a closed tube resonates at a set frequency.

INTRODUCTION

When sound travels through air, air molecules are displaced from their equilibrium positions in a direction parallel to the direction of propagation; sound waves are longitudinal waves. The displacement of the air molecules leads to a variation of the density with alternating regions of compression and expansion. Since the pressure is high where the displacement is low and vice versa, a sound wave may be described in terms of pressure variations as well as in terms of displacement variations. The column of air inside a tube supports most readily sound waves traveling along the axis of the tube. Sound waves are reflected at the end of the tube for both open and closed ends. The sound of musical instruments such as organ pipes, trumpets, and clarinets is due to standing waves in columns of air within a tube.

The resonance condition for a tube depends on whether its ends are open or closed. In this set of experiments, a small speaker at an open end is used to excite vibrations in a tube that may be open or closed at the other end. For a resonant vibration of a tube that is open at both ends, the displacement pattern has antinodes at both ends. For a tube that is closed at one end, the displacement pattern has a node at the closed end and an antinode at the open end, as shown in Figure 1.

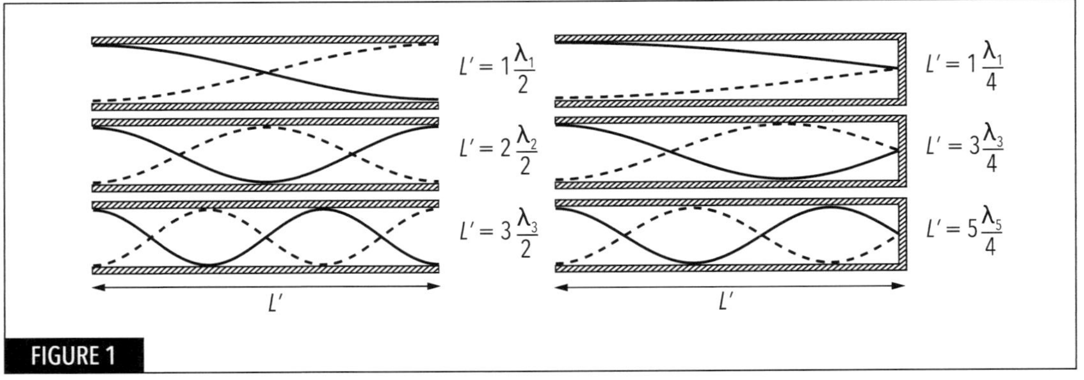

FIGURE 1

Displacement patterns for the three simplest modes of vibration of an open tube (left) and a closed tube (right). The solid and dashed lines show the displacement patterns half a period apart.

The vibrating length, L', of a tube is not exactly equal to its length, L, but has to be corrected for the diameter, d, of the tube. Taking this correction into account, the resonance conditions on the wavelength λ for the open and closed tubes are:

$$L + 0.8d = n\frac{\lambda_n}{2}, \quad n = 1, 2, 3, \dots \text{ open tube} \tag{1}$$

$$L + 0.4d = n\frac{\lambda_n}{4}, \quad n = 1, 3, 5, \dots \text{ closed tube} \tag{2}$$

As for all waves, the wavelength λ is related to the frequency f and the speed v of the wave through

$$v = \lambda f. \tag{3}$$

The speed of sound waves depends on the temperature and is approximately given by

$$v \approx (331 + 0.6T/°\text{C})\text{m/s}, \tag{4}$$

where T is the temperature in degrees Celsius. At room temperature (20 °C), the speed of sound is 343 m/s.

Question 1 For the open tube, show that the resonant frequencies are given by

$$f_n = \frac{v}{2(L + 0.8d)}\, n \tag{5}$$

and derive an equation for the speed of sound from the slope of the graph of f_n as a function of n.

For any given tube length, there are a number of resonant frequencies. Likewise, for a given frequency, there are a number of tube lengths at which standing waves will be formed. To see this, consider the displacement pattern for the fifth harmonic of the closed tube in Figure 1. If you placed a hard wall at the position of either of the nodes, the tube would still resonate with a standing wave of wavelength λ_5. Assuming that you are exciting the closed tube with a fixed frequency f one can show that the resonant lengths L_k are given by

$$L_k = k\,\frac{v}{4f} - 0.4d, \quad k = 1, 3, 5, \dots \tag{6}$$

Question 2 Show that the difference between two consecutive resonant lengths is equal to $\lambda/2$.

LAB EQUIPMENT

- ◎ Resonance tube (length 120 cm, diameter 3.5 cm)
- ◎ Loudspeaker
- ◎ Removable piston
- ◎ Microphone on probe rod
- ◎ Digital function generator
- ◎ Oscilloscope

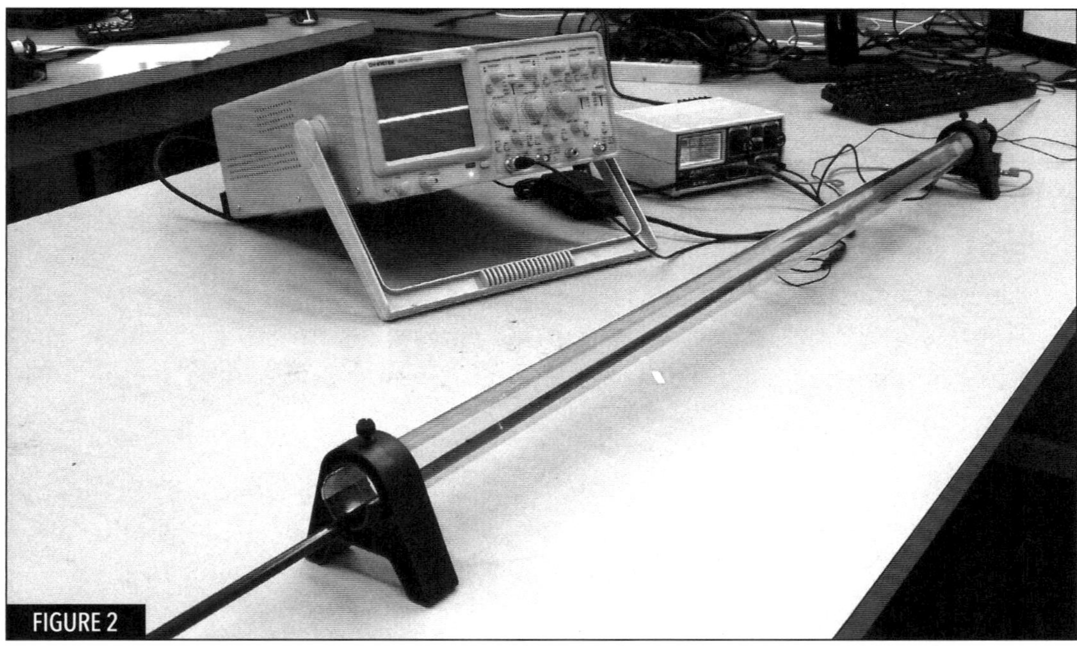

FIGURE 2

Resonance tube with removable piston (inserted from the front end) and probe rod with microphone (inserted from the back end). The speaker is just visible next to the back end of the tube. The digital function generator and oscilloscope are also shown.

Procedure

Connect the speaker to the function generator and the microphone output to the oscilloscope. Insert the probe rod with the microphone some distance into the tube. Place the speaker near the end of the tube where you inserted the microphone. Leave a gap of about one inch between the speaker and the tube.

Part I: Resonances of an Open Tube

Start with a frequency of about 100 Hz and zero amplitude on the function generator. Increase the amplitude slowly until you can just hear a sound from the speaker. Now vary the frequency slowly and listen carefully. In general, the sound will appear louder as the frequency increases, however, you should be able to locate resonances by listening for relative maxima and by observing the trace on the oscilloscope. Find at least six resonances and record the frequency from the display of the function generator along with the order of the harmonic. Plot the resonant frequencies as a function of the order of the harmonic to check if you missed any resonances and perform additional measurements if you did.

Evaluate your data to obtain values for the speed of sound and its uncertainty.

Compare your result with the given value for the speed of sound at room temperature and comment on any discrepancies.

Part II: Resonant Lengths of a Closed Tube

Insert the piston into the tube at the open end opposite the speaker and move it to a position very near the end of the tube. Choose and record a frequency around 900–1100 Hz and adjust the amplitude until you hear the sound clearly. Now push the piston slowly further into the tube until you are near a resonance. Adjust the piston position carefully to make an accurate determination of the resonant length. Record the resonance length and an estimate for its uncertainty and continue moving the piston into the tube until you have found all piston positions that produce standing waves.

Calculate the expected values for the resonance lengths according to Equation (6) and plot these values as a function of k. Add your measured values and their uncertainties to the plot and compare. If you missed a resonance length, go back and try to find it.

Question 3 What is the significance of the slope of the graph?

ELECTROSTATIC INTERACTIONS

OBJECTIVES

- ◎ To observe and interpret electrostatic interactions.
- ◎ To design and carry out an experiment to obtain a rough estimate of the excess charge on a piece of charged tape.

INTRODUCTION

You have all experienced electrostatic interactions, for example, when taking clothes out of a dryer or when combing your hair with a plastic comb on a dry day. Many materials can be electrically charged by rubbing. In the process, electrons are removed from one of the materials and deposited on the other; an example is shown in Figure 1. To "ground" an object means to connect it to Earth with a conductor, such as a conducting wire or a metallic water pipe. The Earth acts like a huge reservoir for charge that can easily accept or provide electrons. Therefore, a charged object loses its excess charge when it is brought in contact with a ground wire or the faucets in the lab.

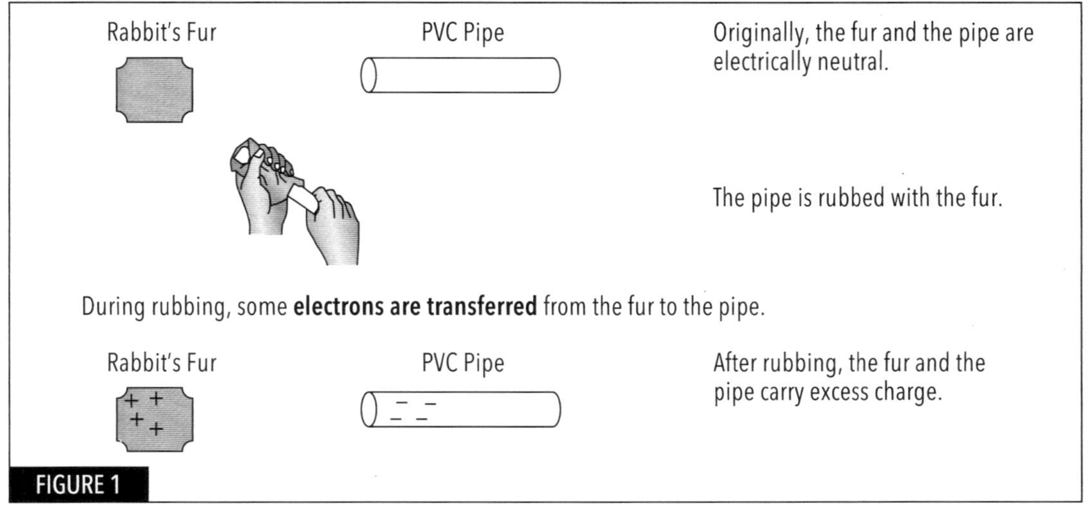

Example of charging by rubbing.

LAB EQUIPMENT

AT EACH TABLE

- Vertical stand with cross bar
- Charge transfer spheres on insulating rods
- Electroscope with Faraday cup
- Ruler

FRONT OF ROOM

- Adhesive tape dispensers
- Plastic rods and pieces of rabbit fur

PROCEDURE

PART I: TAPE EXPERIMENTS

Press a piece of sticky tape, about 6 inches in length, firmly onto a smooth dry surface, for example a plastic notebook or your table. (For ease of handling, make a "handle" by folding one end of the tape back on itself to form a non-sticky portion.) Write a "B" (for bottom) on it. Then press another tape on top of the B tape and label it "T" (for top). Repeat for a second pair of tapes.

Pull each pair of tapes off the table as a unit. While still together, discharge the set by touching its length to a grounded object, for example a faucet. Separate the T from the B tapes and hang one of the B tapes and one of the T tapes from a cross bar. To study the interactions between the hanging tapes and other objects, bring the object near the hanging tape (avoid contact).

For the following interactions, describe your observations of the force: Is it attractive, repulsive, or zero? How does it vary with distance? Please organize your observations in a table.

1. Two T tapes
2. Two B tapes
3. A T and a B tape
4. A T tape and a negatively charged plastic rod (PVC pipe rubbed with rabbit fur)
5. A B tape and the same charged rod
6. A T tape and a grounded object (faucet)
7. A B tape and the same grounded object

Question 1   What can you conclude about the sign of the charge on the T tape and the B tape? What about the magnitude of the charges? Give reasons for your answers and draw sketches of the tapes before and after separation.

Question 2 Explain the interaction between the T tape and the grounded object (faucet). Draw sketches of the tape and faucet when they are far away from each other and when the tape is close by. Then draw the corresponding sketches for the B tape.

Part II: Electroscope Experiments

An electroscope is a device to detect electric charges. The electroscope in this experiment has a hinged metal needle that is free to rotate about its pivot and makes electrical contact with the fixed metal rod but not with the insulated stand. A Faraday cup (a metal container) at the top of the electroscope also makes electrical contact with the metal rod and needle. When the electroscope is uncharged the needle is vertical.

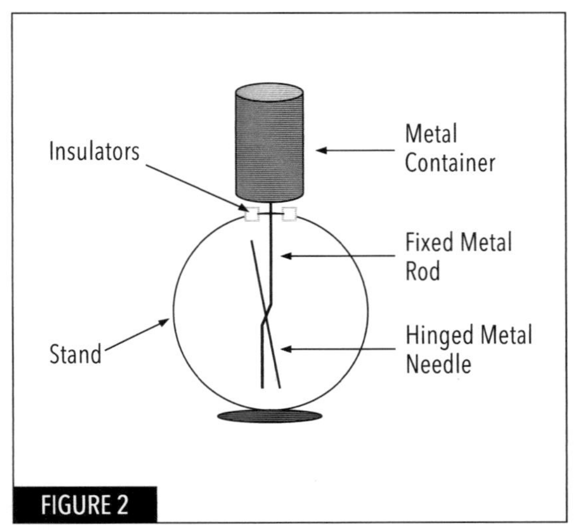

Electroscope with Faraday Cup.

1. Observe the needle when you place a negatively charged plastic rod (rubbed PVC pipe) inside the Faraday cup without touching the metal.

Question 3 Describe the interaction between the charged rod and the electroscope with the Faraday cup. Draw sketches of the rod and electroscope when they are far away from each other and when the rod is inside the cup, and use this to explain why the needle is deflected.

2. To charge the Faraday cup by induction, perform the following steps in order.

   a. Ground the electroscope, by touching it with your hand or a grounded wire, to remove any excess charges.

   b. Place a negatively charged plastic rod inside the Faraday cup.

   c. With the rod inside the cup ground the electroscope with your hand or a wire, making sure that the needle is vertical when you are done.

   d. With the rod inside the cup remove the connection to ground.

   e. Remove the rod and observe the electroscope.

Question 4   What is the sign of the charge (positive or negative) on the Faraday cup after Step 2e? Draw sketches, showing where charges are located at each step, to explain your reasoning.

3. To observe charging by conduction, we will use charged tape as an "electroscope."
    a. Prepare charged tapes as in Part I and suspend them from a support.
    b. Charge the Faraday cup by induction and touch a metal sphere on an insulating rod to the **outside** of the Faraday cup.
    c. Observe the tape strips when you bring the metal sphere close to the strips.

Question 5   What do you conclude about the sign of the charge on the sphere? Is this consistent with your answer to Question 4?

4. Gauss's law implies that if an isolated conductor carries a net charge, the net charge resides entirely on its surface. The following experiment is a test of this principle.
    a. Prepare charged tapes as in Part I and suspend them from a support.
    b. Charge the Faraday cup by induction and touch a metal sphere on an insulating rod to the **inside** of the Faraday cup.
    c. Observe the tape strips when you bring the metal sphere close to the strips.

PART III: ESTIMATE FOR THE EXCESS CHARGE ON A CHARGED TAPE

Design and carry out an experiment to obtain a rough estimate for the amount of excess charge on a charged tape. Here are some hints:

1. The tape has a surface mass density of about 0.067 kg/m^2.
2. For a rough estimate, treat a strip of tape as a small charged ball or point charge.
3. According to Coulomb's law, the magnitude of the force F between two point charges Q_1 and Q_2 at a distance r is

$$F = \frac{1}{4\pi\varepsilon_0}\frac{|Q_1 Q_2|}{r^2},$$

   with $\varepsilon_0 = 8.9 \times 10^{-12}$ C^2/Nm^2.

4. The acceleration due to gravity is

$$g = 9.8 \text{ m/s}^2.$$

Question 6   What is your experiment to determine the excess charge on a piece of charged tape? Draw a sketch of your experimental setup including all quantities that you measure.

Perform the experiment and obtain a value for the excess charge on the tape and for the charge per unit area. The following questions will help you find out if your results makes sense.

Question 7   What is the smallest amount of excess charge a charged tape can possibly carry?

Question 8   What is the largest amount of excess charge per unit area a charged tape can carry in air?

Hints: *When the electric field in air is larger than about 3×10^6 N/C, dielectric breakdown occurs (sparks fly, charges start to flow); the electric field E near an infinite plane of charge is related to the surface charge density σ through $E = \sigma/2\varepsilon_0$.*

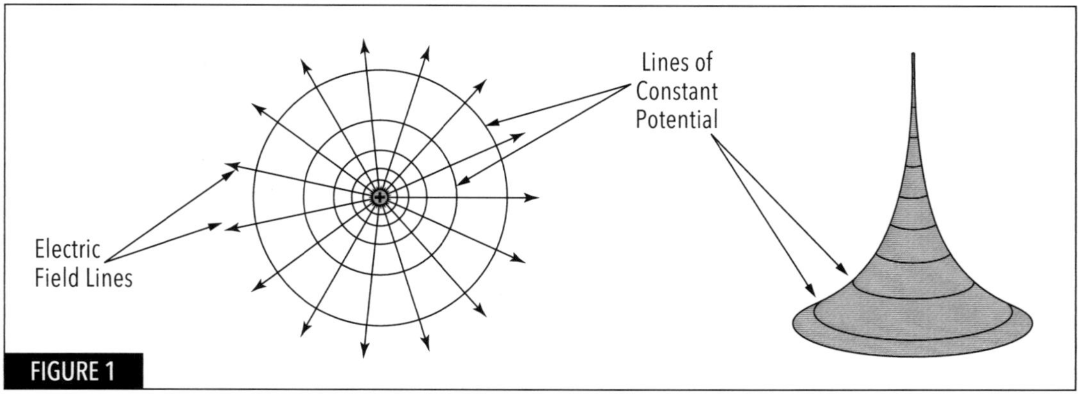

ELECTRIC POTENTIAL AND ELECTRIC FIELDS

OBJECTIVES

- To map the equipotential lines for different electrode configurations representing charged conductors in two dimensions.

- To construct the electric field from the potential map.

INTRODUCTION

Illustration of the electric field and electric potential of a positive point charge in two dimensions. In the left figure, the solid circle at the center (+) represents the charge, the arrows radiating out indicate the electric field, and the concentric circles are lines of constant potential (equipotentials). The figure on the right shows the electric potential of the charge as a landscape; for each point in the x-y plane, the height of the surface indicates the electric potential $V(x,y)$. The lines on the surface are the equipotential lines that are represented by concentric circles in the figure on the left.

Electric field lines and maps of lines of constant electric potential (equipotential lines) are an excellent way to visualize the effect of charges on their environment. Figure 1 shows an example of electric field lines and equipotentials for a positive point charge Q in two dimensions. From Coulomb's law you know that there is a repulsive force between two positive charges. A positive test charge q experiences a force that is directed away from the charge Q. At any point in space, we can find the electric field $\vec{E}$ from the force $\vec{F}$ on the charge, $\vec{F} = q\vec{E}$. The electric field can be visualized with electric field lines; the direction of the electric field lines is the direction of the electric force on a test charge, and the lines are closer together where the field is stronger. Therefore the electric field lines radiate out from the positive point charge in Figure 1. If we want to bring the test charge q closer to the charge Q, we have to do work and increase the electric potential energy U. Since the electric potential V is related to the potential energy of the test charge through $U = qV$, the potential increases as the distance to Q decreases. This is shown in the right part of Figure 1, where the potential is represented by the gray surface, which grows rapidly in the vicinity of the point charge. Since the potential depends only on the distance to Q, the potential is constant on concentric circles.

For conservative forces, such as the Coulomb force between electric charges, the potential energy difference between any two points can be calculated from an integral over the force along a path connecting the points. Similarly, the electric potential difference between any two points can be calculated from an integral over the electric field. If we choose one point (say $\vec{a}$) as a reference, we can calculate the potential V at any point $\vec{r}$ from the equation

$$V(\vec{r}) = - \int_{\vec{a}}^{\vec{r}} \vec{E} \cdot d\vec{l} + V_a, \tag{1}$$

where V_a is the value of the potential at the reference point and $d\vec{l}$ is a segment of the path from $\vec{a}$ to $\vec{r}$. The reference potential V_a is often set to zero for convenience. Equation (1) shows how to calculate the potential by integrating over the electric field. Conversely, the electric field can be calculated from the negative gradient of the potential

$$E_x = -\frac{\partial V}{\partial x}, \quad E_y = -\frac{\partial V}{\partial y}, \quad E_z = -\frac{\partial V}{\partial z}, \tag{2}$$

The negative gradient points in the direction of the steepest decline; therefore, the electric field always points downhill in the potential landscape, and electric field lines are always perpendicular to lines of constant potential.

Electric fields and electric potentials have special properties inside and near the surface of (perfect) conductors in electrostatic equilibrium. (In electrostatic equilibrium, the net force on every particle in the conductor is zero and there is no motion of charges within the conductor.)

The following list is a summary of some of these properties that will help you map the lines of constant potential and construct the electric field lines.

1. Electric field lines start and end at charges or at infinity.

2. If an isolated conductor carries a net charge, the net charge resides entirely on its surface.

3. The electric field is zero everywhere inside a conductor; the electric potential is constant throughout a conductor.

4. The electric field just outside a conductor is perpendicular to the conductor surface; the equipotentials just outside a conductor are parallel to the surface of the conductor.

5. For an irregularly shaped conductor, the electric field is highest where the radius of curvature of the surface is smallest, for example near sharp tips.

LAB EQUIPMENT

- ◎ Overbeck apparatus
- ◎ Plates with electrodes (parallel lines, dipole, point and line, Faraday cup)
- ◎ Stencils for tracing patterns
- ◎ PASCO® interface and software
- ◎ Voltmeter
- ◎ Banana plug patch cords and alligator clips

PROCEDURE

OVERBECK APPARATUS SETUP

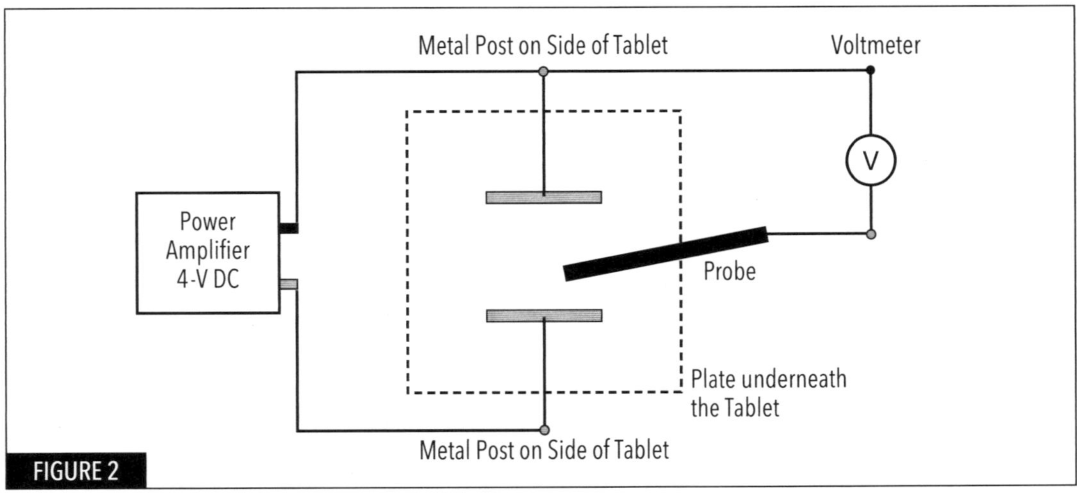

Overbeck apparatus.

PREPARATION

Attach the plate with the parallel lines to the bottom of the tablet and a sheet of paper on top of the tablet. Secure the paper with the rubber stubs and with a piece of tape. Using the appropriate plastic mask, draw an outline of the conductors on the paper and then use a ruler to draw a straight horizontal line through the center of the outlines (this will become your x-axis).

Set up the power amplifier as a 4-V DC power supply as described in the instruction sheet, then connect the wires as shown in Figure 2. Hold the U-probe so that the leg with the hole is above the tablet and the leg with the contact is below the tablet. Position the probe so that it makes contact with one of the conductors on the plate and read the voltmeter. Move the probe to the other conductor and read the voltage again. If everything is set up correctly, you read voltages of 0 V and about 4 V; if not check your setup and ask for help.

MAPPING EQUIPOTENTIALS

Equipotentials are locations of constant potential; in two dimensions, they are lines. To find a particular equipotential line, for example for a potential difference of 3.5 V between one conductor and the probe, move the probe slowly until you find a position where the voltage reads 3.5 V. Mark the position on the paper with a pencil and move the probe until you find another position where the reading is 3.5 V. Repeat this until you have enough points to draw the 3.5-V equipotential line by connecting the points.

1. ***Two Parallel Lines*** (model for a parallel plate capacitor)
   a. Map the equipotentials, for voltages from 0.5 V to 3.5 V in steps of 0.5 V (please be sure to note the voltage on the paper).
   b. Draw the equipotential lines by connecting points of the same potential.
   c. Construct electric field lines (about eight lines emanating from the positive conductor and ending at the negative conductor). Please be sure to include the direction of the electric field lines.
   d. For each voltage, measure the x-position of the equipotential line when it crosses the x-axis. Make a table in Excel and plot the equipotential voltage (in volts) as a function of x (in meters).

    e. Do your data fall on a straight line? If not, check your data and repeat some measurements if necessary.

    f. In Excel, calculate the differences ΔV and Δx for the six pairs of neighboring positions, calculate $E_x = -\Delta V/\Delta x$, and plot E_x as a function of x. If the data make sense, calculate the average and standard deviation of E_x and compare your result for the magnitude of the electric field with that expected for a parallel plate capacitor ($E = V/d$).

2. **Two Points** (electric dipole)

    a. Map the equipotentials for voltages from 0.5 V to 3.5 V in steps of 0.5 V (please be sure to note the voltage on the paper).

    b. Draw the equipotential lines by connecting points of the same potential.

    c. Construct electric field lines (about eight lines emanating from the positive conductor and ending at the negative conductor). Please be sure to include the direction of the electric field lines.

3. **Point and Line**

    a. **Before you start your measurements,** draw an outline of the "point and line" conductor on a clean sheet of paper and draw your predictions for electric field lines and equipotential lines on the paper. Label the page "Prediction" and put it aside.

    b. On a fresh page, map the equipotentials for voltages from 0.5 V to 3.5 V in steps of 0.5 V; then add two more equipotentials with 0.25 V steps near the point conductor if possible (please be sure to note the voltage on the paper).

    c. Draw the equipotential lines and construct electric field lines as before.

    d. Compare your experimental results with your predictions.

4. **Faraday Cup** (U-shaped conductor with point source)

    a. Map the equipotentials for voltages from 0.5 V to 3.5 V in steps of 0.5 V; then add two more equipotentials with 0.25 V steps in the interior of the Faraday cup (U-shape), if possible (please be sure to note the voltage on the paper).

    b. Draw the equipotential lines and construct electric field lines as before.

    c. Identify the regions where the electric field is strongest and where it is weakest.

    d. Remembering that electric field lines start and end on charges or at infinity, what can you say about the charges in the interior of the Faraday cup?

Question 1   You know that the electric field is the negative gradient of the potential. Why does this allow you to calculate the x-component of the electric field (approximately) as $E_x = -\Delta V/\Delta x?$

Question 2   For the parallel lines, are your results for E_x consistent with your electric field line pattern? In particular, from your electric field lines, what do you expect for the sign of E_x? Does this agree with your results for E_x? What can you say about the magnitude of E_x? Does this agree with your results for E_x? What would you expect to find for E_y?

Question 3   The two point conductors form a dipole when a potential difference is applied. How can you tell from your equipotentials which of the points has the positive charge?

Question 4   Can equipotential lines cross? For example, could 3-V and 2-V equipotential lines have a point in common? Explain your answer.

Question 5   Can electric field lines cross? Explain your answer.

RESISTANCE

OBJECTIVES

◎ To test the validity of Ohm's law for resistors on a circuit board and determine their resistance.

◎ To study the effect of the temperature dependence of the resistivity on the current-voltage characteristics of a light bulb.

INTRODUCTION

A current-voltage characteristic or $I\text{–}V$ curve is a graph of the current I through a device as a function of the applied voltage V. Ohmic resistors have very simple $I\text{–}V$ curves since the current through the resistor is proportional to the voltage. As shown in Figure 1, the graph of the current as a function of the voltage is a straight line whose slope is the inverse of the resistance R.

Ohm's law summarizes the relation between voltage, current, and resistance for ohmic resistors:

$$I = \frac{1}{R} V \qquad\qquad \text{Ohm's Law} \quad (1)$$

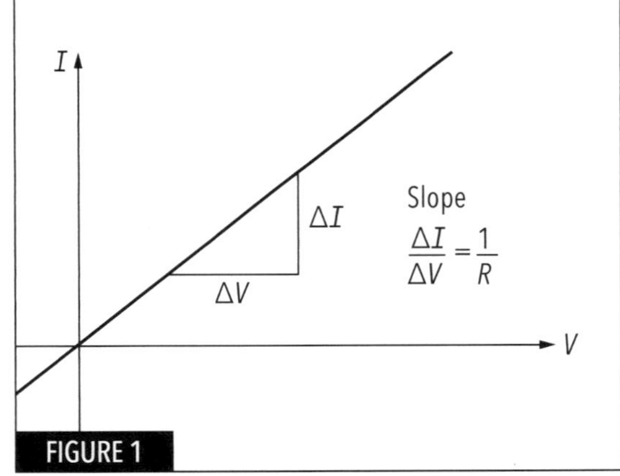

FIGURE 1

Current-voltage characteristic for an ohmic resistor. The graph of the current I as a function of the applied voltage V is a straight line with intercept zero. The slope of the line is the inverse, $1/R$, of the resistance R.

In the SI system of units, the unit of current is ampere (A), the unit of voltage is volt (V), and the unit of resistance is ohm = volt/ampere (Ω = V/A).

The resistance of an ideal ohmic resistor is independent of the size and frequency of the applied voltage. For real resistors, such as metallic wires and resistors that you find on circuit boards, this is true only for a limited range of working conditions. In the first parts of this experiment, you will measure *I–V* curves of resistors on a circuit board, test the validity of Ohm's law, and determine the resistance. In Part III you will investigate a light bulb.

Light bulbs can show very nonohmic *I–V* characteristics. In an incandescent light bulb a thin wire, the filament, is enclosed by a glass bulb under vacuum; see Figure 2 for an illustration. When a current flows, the resistance of the wire causes energy dissipation so that the filament heats up to the point where it glows and emits light. For a cylindrical wire, the resistance is calculated from

$$R = \rho\,\frac{L}{A}, \tag{2}$$

where ρ is the resistivity, L is the length of the wire, and A is the cross-sectional area of the wire. The resistivity of metal wires is due to collisions of conduction electrons in the wire. As the temperature increases, these collisions become more frequent and the resistivity increases. Figure 3 shows a graph of the resistivity of tungsten as function of temperature. The solid line represents reference values from the National Institute of Standards and Technology (NIST) [1]. The dashed line is calculated from the linear equation

$$\rho(T) = \rho_0[1 + \alpha(T - T_0)], \tag{3}$$

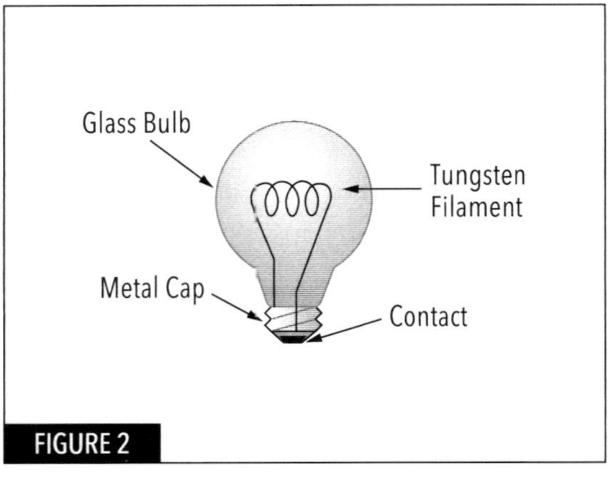

FIGURE 2

Illustration of an incandescent light bulb. The filament of the light bulb is a thin tungsten wire. One lead to the filament makes electrical contact with the metal cap, the other with the contact area at the bottom of the base.

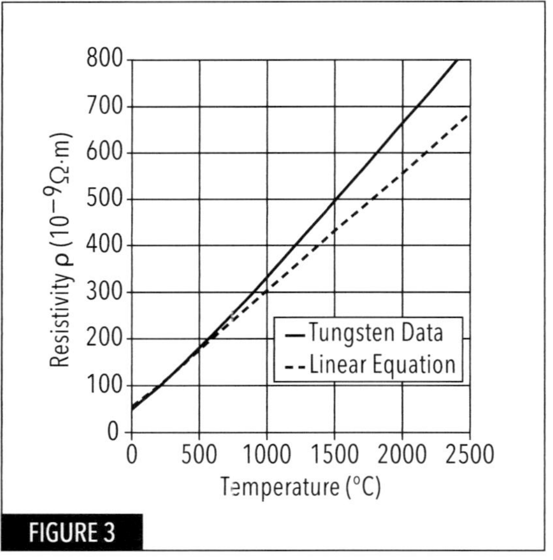

FIGURE 3

Resistivity of tungsten as a function of temperature. The solid line represents reference data [1], the dashed line shows values calculated from the linear Equation (3).

with the parameter values $\rho_0 = 56 \times 10^{-9}$ $\Omega{\cdot}$m, $\alpha = 0.0045$ $(^\circ$C$)^{-1}$, and $T_0 = 20$ $^\circ$C from Ref. [2]. The graphs show that the resistivity increases almost linearly with temperature near room temperature and more strongly at high temperatures. This suggests that a light bulb will not act as an ohmic resistor when the filament heats up from room temperature to the operating temperature, which can be as high as 3000 $^\circ$C.

LAB EQUIPMENT

- ◎ RLC circuit board
- ◎ Voltage probe
- ◎ Banana plug patch cords
- ◎ PASCO interface and software

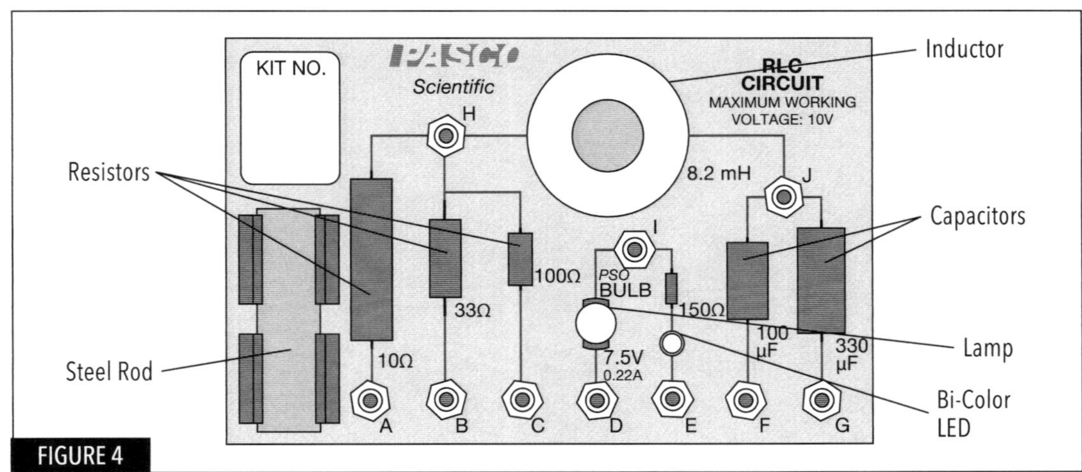

PASCO RLC circuit board. Connectors are labeled by capital letters.

PROCEDURE

PART I: CURRENT THROUGH A RESISTOR R IN RESPONSE TO A CONSTANT APPLIED VOLTAGE

Set up the power amplifier as a DC power supply and set up the voltage probe as described in the instruction sheet. Create two graphs, one for the voltage measured by the probe as a function of time, the second for the current output from the power amplifier as a function of time. Connect the output from the power amplifier and the leads from the voltage probe according to the diagram in Figure 5. Here R is the resistor labeled 33 Ω on the circuit board, this is the resistor between connectors B and H in Figure 4.

1. Apply a 5-V DC voltage across the resistor, and measure the voltage and the current. After a few seconds, step down the voltage to 4, 3, 2, and 1 V leaving a few seconds in between to allow the reading to stabilize. Determine the voltage and current values for each setting and transfer the data to Excel. Plot the current as a function of voltage.

Question 1 Does the I–V graph support that R is an ohmic resistor? If yes, determine the resistance and compare with the printed value. If not, check your measurements.

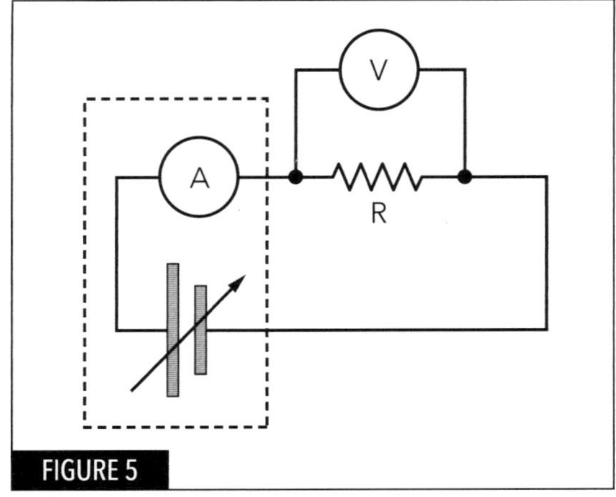

FIGURE 5

Circuit diagram for measuring the current through a resistor R in response to an applied DC voltage. The box of dashed lines indicates the components of the power amplifier; an adjustable voltage source (battery symbol with arrow) and an integrated ammeter (A). The voltage probe (V) measures the voltage across the resistor R.

PART II: I–V CURVE FOR TIME-VARYING APPLIED VOLTAGE

1. Set the power amplifier output to a triangular wave of amplitude 2.5 V and frequency 0.01 Hz and create a graph with the current as a function of probe voltage. Measure the I–V curve over one period (100 s). If the I–V curve supports that R is an ohmic resistor, determine the resistance and its uncertainty from the slope and compare with the value from Part I. If not, or if the value is significantly different from the first value, check your measurements.

2. Repeat the measurement and evaluation for two more frequencies, 1 Hz and 100 Hz. For each frequency, adjust the sampling rate to avoid data overflow at higher frequencies. It is a good idea to set an automatic stop time of one period, $T = 1/f$.

Collect all results for this resistor in a table and discuss if the resistor is ohmic for the range of applied voltages and frequencies you investigated. Also comment on the accuracy of the resistance value printed on the circuit board.

Question 2 How will other sources of resistance (leads) affect the results of your resistance measurements?

Part III: *I–V* Characteristics of a Light Bulb

Question 3   What kind of *I–V* curve do you expect when the voltage across a light bulb is increased extremely slowly? Draw a sketch and give a brief explanation.

1. Change the setup so that you can measure the *I–V* curve for the light bulb on the board. Set the power amplifier output to a triangular wave with a 4.5-V amplitude. You will perform measurements at three frequencies, one high frequency (1000 Hz), one intermediate frequency (1.0 Hz) and one very low frequency (0.001 Hz). Please don't forget to adjust the sampling frequency and time. For each frequency record the *I–V* curve and observe how the brightness of the light bulb changes with time.

2. As you go along, study the *I–V* curves and decide if they show ohmic behavior for the whole curve, for voltages close to zero only, or if they show "hysteresis," i.e., the *I–V* curve for decreasing voltage does not retrace the curve for increasing voltage. If you find ohmic behavior determine the resistance by fitting to the appropriate part of the curve. If you find a hysteresis loop, note which branch of the loop corresponds to increasing and which branch to decreasing voltage.

Make a table of your observations about the graphs, the resistance, and the brightness variation of the bulbs. For low, intermediate, and high frequencies interpret the current-voltage characteristic in terms of increasing, decreasing, or steady temperature and resistance.

Describe a method to find a value of the temperature of the glowing filament from the measurements in this experiment and apply it to obtain an estimate for the temperature. ***Hint:*** *Use your high frequency results to find the temperature and do the calculations while you perform the experiment at very low frequency.*

Question 4   When is it more likely for a light bulb to burn out, when it is first switched on or after it has been burning for a while? What is the reason for this behavior?

References ──

1. J.G. Hust and P.J. Giarratano, Standard Reference Material 799, *Electrical Resistivity—Tungsten*, Nat. Bur. Stand. Special Publication 260-52 (1975). *http://ts.nist.gov/MeasurementServices/ReferenceMaterials/archived_certificates/799.pdf*

2. Douglas C. Giancoli, *Physics for Scientists and Engineers*, 4th edition, Pearson, Upper Saddle River, NJ, 2008.

RC CIRCUITS

OBJECTIVES

- ◎ To measure the voltage across a capacitor during charging and discharging.
- ◎ To determine the capacitance from the time constant and the resistance.
- ◎ To construct a simple capacitor and determine its capacitance

INTRODUCTION

Capacitors are circuit elements designed to store charge and energy. They consist of two conductors separated by a dielectric. When a voltage V is applied across a capacitor, one of the conductors acquires a positive charge $+Q$ and the other a negative charge of the same size, $-Q$. The charge on the conductors is proportional to the voltage and the constant of proportionality is the capacitance C:

$$Q = CV \tag{1}$$

A simple capacitor consists of two metal sheets separated by a dielectric material. The capacitance C of a parallel plate capacitor is given by

$$C_{\text{plate}} = K\varepsilon_0 \frac{A}{d}, \tag{2}$$

where $\varepsilon_0 = 8.85 \times 10^{-12}$ C/(V·m) is the permittivity of free space, K is the dielectric constant of the material, A is the area of the plates and d is the distance between the plates. To save space and protect the capacitor from moisture, many commercial capacitors contain an extra layer of dielectric and are rolled up tightly and sealed. For the same area and distance between the layers, the capacitance is larger since charges can now be stored on both sides of the conductor

$$C_{\text{roll}} = \left(2 - \frac{1}{n}\right) K\varepsilon_0 \frac{A}{d}, \tag{3}$$

where n is the number of turns of the capacitor. In the last part of the experiment, you are going to construct a capacitor, estimate and measure its capacitance, and explore how squeezing the layers changes the capacitance.

The rules for calculating the equivalent capacitance of a capacitor combination are similar (but not identical!) to those for calculating the equivalent resistance of a resistor combination. Table 1 summarizes the equations for combinations of two capacitors, C_1 and C_2, and of two resistors, R_1 and R_2.

TABLE 1

Equivalent capacitances of combinations of two capacitors, C_1 and C_2, and equivalent resistances of combinations of two resistors, R_1 and R_2.

CONNECTED IN PARALLEL	CONNECTED IN SERIES
$C_{eq} = C_1 + C_2$	$C_{eq} = \dfrac{C_1 C_2}{C_1 + C_2}$
$R_{eq} = \dfrac{R_1 R_2}{R_1 + R_2}$	$R_{ec} = R_1 + R_2$

In the first part of the experiment, you will learn how to determine the capacitance by measuring the voltage across the capacitor as a function of time. Figure 1 shows a circuit diagram for charging and discharging a capacitor through a resistor.

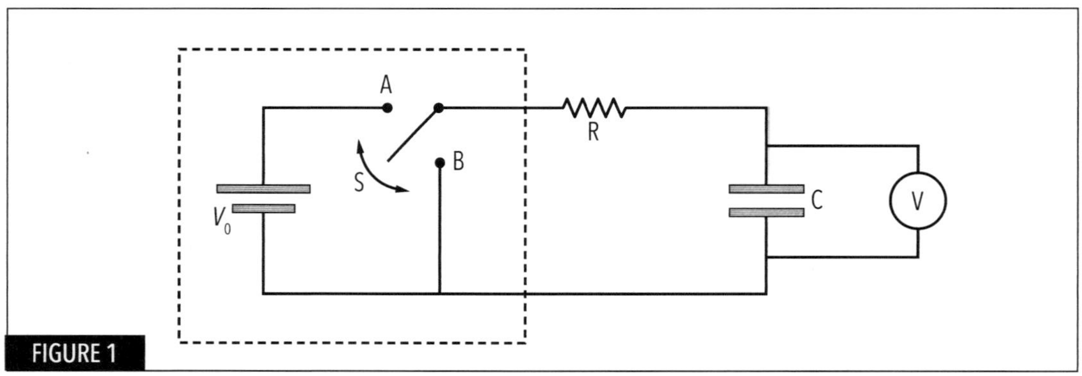

FIGURE 1

Circuit diagram for measuring the voltage across a capacitor C during charging and discharging through a resistor R. When the switch (S) is in position A, the battery voltage V_0 is applied across the resistor and the capacitor. When the switch is in position B, the voltage source is disconnected. The voltage probe (V) is set up to measure the voltage across the capacitor C. In our experimental setup, the power amplifier and signal generator play the role of the battery and the switch; the dashed lines enclose the corresponding components.

We start with an uncharged capacitor and apply a voltage V_0 across the resistor and capacitor at time $t = 0$ by moving the switch S to position A. Initially, the charge on the capacitor increases rapidly, but then it levels off and reaches a constant value Q_0 at very long times. The graph of $Q(t)$ in the left panel of Figure 2 illustrates this behavior. The equation for the charge on the capacitor during charging is

$$Q(t) = Q_0(1 - e^{-t/\tau}), \qquad\qquad\qquad \text{Charging} \quad (4)$$

where

$$\tau = RC \qquad\qquad\qquad (5)$$

is the time constant of the RC circuit (RC time). If we start at time $t = 0$ with a charged capacitor and move the switch to the B position, the capacitor discharges through the resistor. The charge on the capacitor as a function of time is a decreasing exponential with the same time constant, $\tau = RC$, as during charging,

$$Q(t) = Q_0 e^{-t/\tau}. \qquad\qquad\qquad \text{Discharging} \quad (6)$$

The right panel of Figure 2 shows a graph of the charge on the capacitor as function time during discharging.

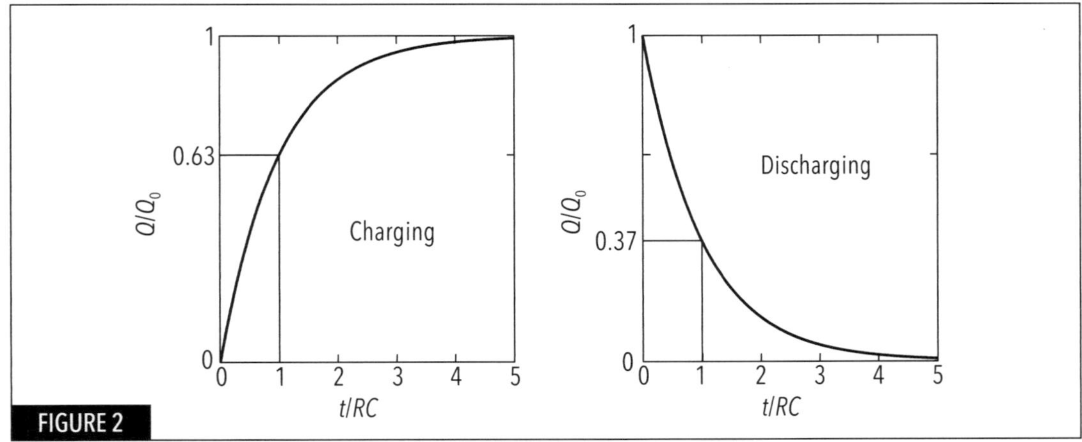

FIGURE 2

Charge on a capacitor during charging (left panel) and discharging (right panel). The graphs show the charge on the capacitor Q divided by the maximum charge Q_0 as a function of time t divided by the time constant $\tau = RC$. During charging, at time $t = RC$ the charge on the capacitor has reached $(1 - 1/e) = 0.63$ of its maximum value; during discharging, at time $t = RC$ the charge has dropped to $(1/e) = 0.37$ of its maximum value.

In the experiment, the power amplifier and signal generator are set to produce a positive square wave signal of a suitable frequency f. This corresponds to having the switch in position A for half of the period $T = 1/f$ and having the switch in position B for the other half of the period. In the experiment you will measure the charge indirectly by measuring the voltage across the capacitor. Your voltage signal as a function of time should look similar to the graph in Figure 3.

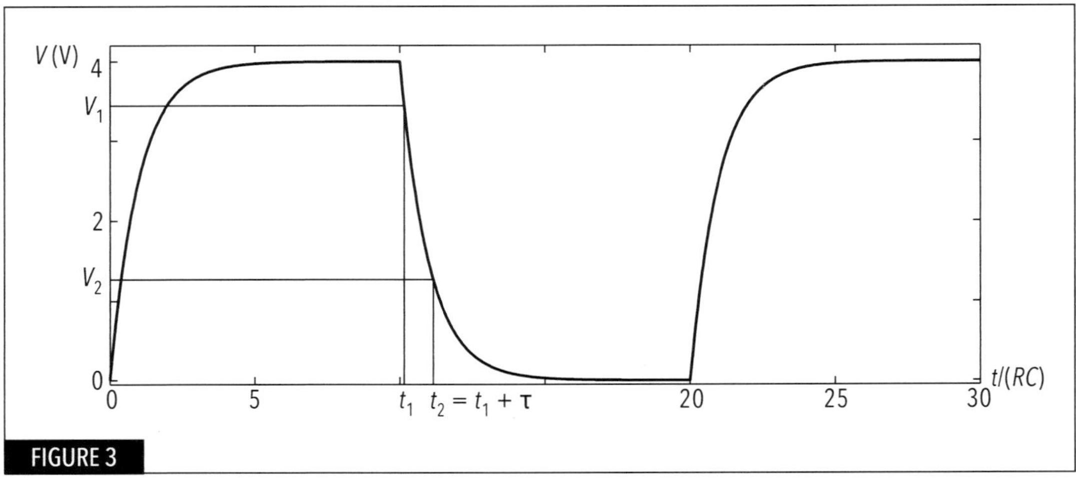

FIGURE 3

Voltage across the capacitor during charging and discharging with a positive square-wave signal of amplitude 4 V and frequency $f = 1/(20\,RC)$. The voltage is shown as a function of time divided by the time constant for a time of 3/2 signal periods. The times and voltages indicated by the thin lines pertain to Question 2.

LAB EQUIPMENT

- RLC circuit board
- Voltage probe
- Rolls of aluminum foil and wax paper
- 1000 Ω resistor
- Banana plug patch cords
- PASCO interface and software
- Scissors and tape
- PVC pipe
- Alligator clips

PROCEDURE

PART I: TIME CONSTANT OF AN RC CIRCUIT

1. Choose either the 33 Ω or the 100 Ω resistor and one of the capacitors on the RLC circuit board and record the printed values for R and C. Connect the power amplifier and the voltage probe according to the circuit diagram in Figure 1. Also put a jumper cable across the inductor (copper coil) on the circuit board. Set up the power amplifier to provide a positive square wave signal of amplitude 4 V and a frequency of $f = 1/(20RC)$. Set the sampling rate to 2000 times the signal frequency, if possible, but do not exceed 25,000 Hz, and stop measurements after two signal periods. Resize the figure appropriately before saving and printing it.

You are going to use two independent methods to determine the time constant ($\tau = RC$) from the measured $V(t)$ curve. First, you will find τ by evaluating two data points and then you will use software to fit the data. If both results agree within their uncertainties, you will have confidence in your results and you may use your preferred method for the remaining experiments.

Question 1   What are the equations for the voltage across the capacitor as a function of time? Use Equations (4) and (6) together with Equation (1) to find $V(t)$.

Question 2   Do you need the data points exactly at time $t = 0$ to extract the time constant from the graphs? Use your equations and refer to Figure 3 to show the following:

During **_discharging_**, if the voltage across the capacitor has a value of V_1 at time t_1, the voltage V_2 at time $t_2 = t_1 + \tau$ is given by

$$V_2 = e^{-1} V_1 \cong 0.368V_1. \tag{7}$$

How does this help you to determine the time constant? Draw sketches and indicate what you need to measure to find the time constant from the discharging portion of the $V(t)$ curves.

2. Identify the time intervals on the graph where the capacitor was charging and where it was discharging; mark these time intervals on your printout. Use the cross hair tool in Capstone to find values for V_1, V_2, t_1, t_2 and determine the time constant during discharging from Equation (7). Estimate the uncertainty for the RC time obtained by this method.

3. Now use the software fitting tool to obtain a value for the RC time. Highlight a section of your $V(t)$ graph that corresponds to discharging the capacitor and perform a Natural Exponent Fit. If the fitted curve agrees well with your experimental data, record the fit parameters and save the graph. If the fitted curve deviates significantly from your data, change the range of data that is being fitted until you have a good fit. Keeping in mind that the fitting tool assigns the time $t = 0$ to the first selected data point, compare the fit equation with your equation for the voltage during discharging and determine the RC time from the fit. Please be sure to include an uncertainty estimate.

4. Collect your time-constant data for the RC combination (call them R_1 and C_1) in a spreadsheet. Do the results agree within their uncertainties? If not, discuss what the problem may be and repeat a measurement or revise an error estimate, as appropriate.

5. Determine the capacitance and its uncertainty using the known value of the resistance.

Part II: RC Combinations

1. Use your preferred method to determine the RC time for the same capacitor (C_1) and a different resistor (R_2).

2. Use your preferred method to determine the RC time for the same capacitor (C_1) and the parallel combination of R_1 and R_2 (you will need to add a jumper cable for this).

3. Organize your data for the RC times and determine the capacitance using the known values of the resistances. Do the results agree within their uncertainties? If yes, find the average and standard deviation. If not, discuss what the problem may be (see also Question 3) and repeat a measurement or revise an error estimate, as appropriate.

4. Compare your value for the capacitance with the value printed on the circuit board. According to the manufacturer of the boards (PASCO) the tolerance of the capacitance values is ±20%. Do your values agree within the tolerance?

Question 3 What are the most important uncertainties in your results for the capacitance? Are there other resistances that should have been taken into account? How would you expect the capacitance values from the individual resistors and the parallel combination to be affected by such resistances?

Question 4 Which capacitor/resistor combination of this circuit board allows you to store the largest amount of charge with the largest time constant? Study the circuit board and draw a circuit diagram for a circuit that can be realized with this board and as many jumper cables as needed. Calculate the RC-time and the amount of charge stored, assuming the output voltage of the power amplifier is 4.0 V.

Question 5 Repeat Question 4 for the capacitor/resistor combination with the *largest amount of charge* and the *smallest time constant.*

5. Perform experiments to determine the capacitance of a second capacitor (C_2).

6. Test your predictions for the RC times in Questions 4 and 5 experimentally.

Question 6   The following warning is found in a car-repair manual: "Disconnect and isolate the battery negative (ground) cable, then wait two minutes for the system capacitor to discharge before performing further diagnosis or service." What is the reason for this warning? How might the suggested waiting time be related to the *RC* time of the circuit? How would the discharge time change if something (for example a screwdriver) with a substantially smaller resistance were to touch both leads of the capacitor?

PART III: BUILD YOUR OWN CAPACITOR

In this part of the experiment, you are going to roll your own capacitor. Figure 4 shows illustrations of the layers that form the capacitor, seen from the top, from the side, and of the capacitor as it is being rolled up.

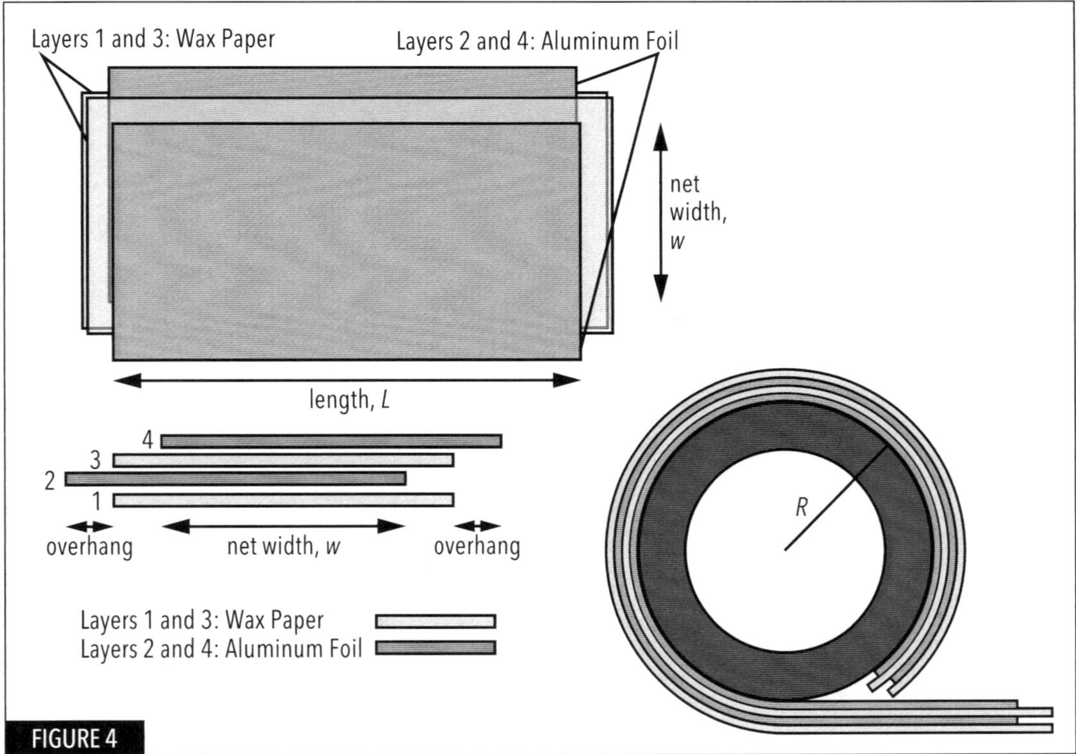

Top Left: Top view of the layers before the capacitor is rolled up. Layer 1 (bottom layer) and layer 3 are wax paper and aligned with each other. Layers 2 and 4 are aluminum foil layers and a little shorter than the wax paper. Aluminum layer 2 is shifted up and layer 4 is shifted down relative to the paper layers. The length L is the length of the aluminum sheets, the net width w is the width of the foil minus the overhangs on either side of the paper layers; the area of the capacitor is calculated from A = Lw. **Bottom Left: View of the layers from the left side. Bottom Right: Side view of the capacitor while the layers are being rolled onto a core (PVC pipe) of radius R.**

Question 7   To obtain a first (ideal) estimate for the capacitance, evaluate Eq. (3) assuming that the separation between the aluminum sheets is equal to the thickness of the paper (no air layers). The dielectric constant of wax paper is $K =$ 2.5, the thickness is $d = 2.8 \times 10^{-5}$ m, the area is $A = Lw$, and the number of turns is $n = L/(2\pi R)$.

Question 8   For a second estimate, we assume that a thin layer of air is trapped next to the wax paper layers. This is equivalent to a second capacitor, with dielectric constant $K = 1$ and thickness d_{air}, connected in series with the wax paper capacitor. Assume that $d_{air} = d$ and calculate the capacitance of the air capacitor, C_{air}, from Eq. (7). Then calculate the equivalent capacitance for the rolled capacitor with air from the equation for capacitors in series in Table 1. You should find that the equivalent capacitance is smaller than both the ideal capacitance and the air-filled capacitance.

1. Cut two sheets of wax paper, about 80 cm in length, and two sheets of aluminum foil, about 75 cm in length, and arrange the layers flat on the table as shown in the top left of Figure 4. Record the radius R of the PVC pipe, the length L and the effective width w.

2. Place the PVC pipe at the left edge of the assembly, hold on tightly to the right edge of the layers, and start rolling up the capacitor as tightly as you can. Secure the roll with two strips of masking tape. You should have an outer layer of wax paper with aluminum foil extending beyond the paper on both sides. Fold up a corner of the top layer of aluminum on either side of the paper; this will serve as your contact in the RC circuit.

3. Connect your capacitor in series to the 1000 Ω resistor and the power amplifier as shown in Figure 1; connect the voltage probe so that it measures the voltage across your capacitor. Set up the power amplifier to provide a positive square wave signal of amplitude 4 V and frequency 500 Hz. Set the sampling rate to 100,000 Hz and stop measurements after two signal periods. Determine the RC time of the circuit and calculate the capacitance and its uncertainty. Compare your results with the estimates from Questions 7 and 8.

4. The capacitance is very sensitive to the distance between the conducting sheets. Demonstrate this effect by taking measurements while you squeeze the middle of the capacitor roll with your hands. Determine the capacitance and compare with your previous results.

MAGNETIC EFFECTS

OBJECTIVES

◎ To measure the magnetic field produced by the current through an air-filled coil.

◎ To investigate the superposition of magnetic fields.

◎ To observe electromagnetic induction.

INTRODUCTION

You can investigate the magnetic field surrounding a permanent magnet with a compass needle. The needle itself is a small permanent magnet and aligns itself so that its north pole points to the south pole of the magnet. The magnetic field can be visualized with magnetic field lines; the direction of the magnetic field lines is the direction of the compass needle at that point, and the lines are closer together where the field is stronger. Moving charges, such as the electrons flowing through a current carrying wire, also produce magnetic fields. If there is more than one source of magnetic fields, the net magnetic field $\vec{B}$ at any point is the vector sum of the magnetic fields $\vec{B}_1, \vec{B}_2, \ldots$ due to the different sources.

$$\vec{B} = \vec{B}_1 + \vec{B}_2 + \ldots \qquad \text{Superposition of Magnetic Fields} \quad (1)$$

A current carrying wire is surrounded by a magnetic field whose strength is proportional to the current and whose field lines form closed circles with the wire at the center. When the wire is wound into a coil, the magnetic field at any point is the vector sum of the magnetic fields due to the turns of the coil. For a very long and thin coil (solenoid) the magnetic fields between neighboring turns cancel each other, while inside the coil, the magnetic fields add up to a net field that is parallel to the axis of the coil. Near the ends of the coil, the magnetic field lines spread out and the field is no longer uniform. In this lab, you will not use solenoids but large air-filled wire coils. The diameter of the coils is approximately equal to their length, so that the end-effects on the magnetic field extend some distance into the coils. Figure 1 shows a photograph of such coils and an illustration of the magnetic field. When more than one current-carrying coil is present the magnetic fields due to the individual coils add up to give the total field (superposition).

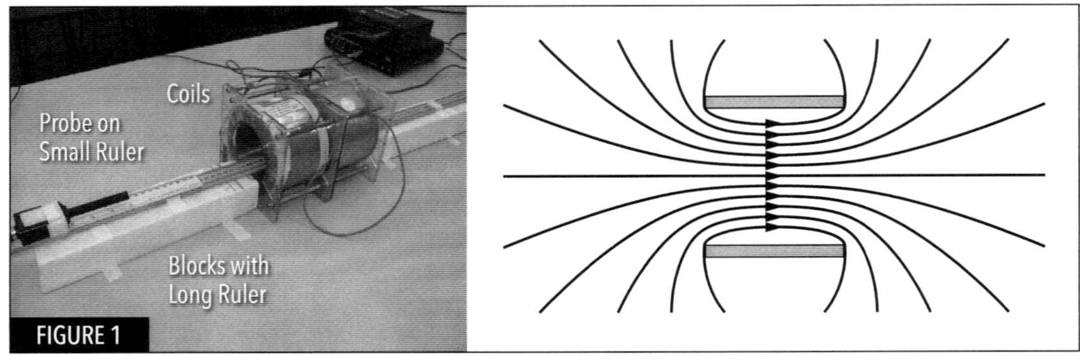

Setup for Part II of the experiment and illustration of the magnetic field in a cross section through the axis of a coil. The lines with arrows indicate magnetic field lines and the solid rectangles the cross sections of the wire layers.

While a steady current produces a steady magnetic field, a steady magnetic field does not produce a current. However, if you move a permanent magnet into or out of a coil then a current is induced in the coil. This phenomenon is called electromagnetic induction and is due to the magnetic flux through the coil changing in time. A good way to visualize the magnetic flux through the coil is to count the number of magnetic field lines that penetrate it. The magnetic field of a permanent magnet is strongest near the poles so that the field lines are closest together there. If you bring the magnet closer to the coil, more of its field lines will pass through the coil thus changing the flux through the coil and inducing a current. Instead of moving a permanent magnet you can also induce a current in the coil by changing the magnetic field produced by a second coil nearby. You will observe electromagnetic induction in the third part of this experiment.

Lab Equipment

- ◎ Meter stick and ruler
- ◎ Pink Styrofoam™ board
- ◎ Banana plug patch cords
- ◎ Current sensor

- ◎ Bar magnet
- ◎ PASCO interface and software
- ◎ 2 coils
- ◎ Magnetic field sensor (Hall probe)

In all experiments, you will measure magnetic fields with a Hall (effect) probe.
Figure 2 shows drawings of a Hall probe and its controls.

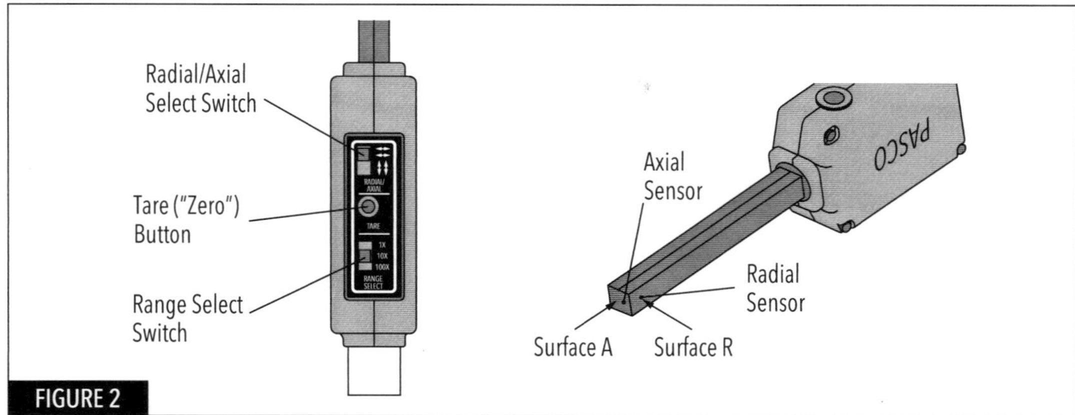

Hall (effect) probe to measure magnetic fields. The drawing on the left shows the controls; the
drawing on the right shows the location of the sensors.

When the Radial/Axial Select Switch on the Hall probe is set to:

◎ Axial, the component of the magnetic field perpendicular to surface A
is measured.

◎ Radial, the component of the magnetic field perpendicular to surface R is
measured.

The arrows next to the radial/axial switch indicate the direction that is counted
as positive. Since Hall probes are very sensitive to position and orientation, keep
the probe as still as possible during measurements. The tare button sets the reference of the magnetic field; when you press the tare button, the present magnetic field is subtracted from future measurements. This allows you to subtract
constant fields such as the earth's magnetic field. Make sure you always "tare" the
probe between measurements.

Procedure

Tape the Hall probe to the ruler, set the switch on the probe to axial, and set up
the software for the power amplifier and the magnetic field sensor, setting the
range switch on the probe to 10×.

PART I: DEPENDENCE OF THE MAGNETIC FIELD STRENGTH ON THE CURRENT THROUGH THE COIL

Question 1 How do you expect the magnetic field to depend on the current through the coil?

1. Connect a coil to the power amplifier output and set it up to generate a triangular wave of amplitude 10 V and frequency 0.01 Hz with sampling frequency 20 Hz. Create a graph of the magnetic field as a function of the current through the coil. Tare the Hall probe, place the ruler with the probe so that it is supported by Styrofoam blocks and the probe is in the center of the coil. Take measurements for one period and resize the graph appropriately before saving it. Describe how the magnetic field depends on the current and compare with your prediction.

Question 2 What do you expect to happen when you reverse the direction of the current, for example by exchanging the connecting wires?

PART II: SUPERPOSITION OF MAGNETIC FIELDS

Question 3 How do you expect the strength of the magnetic field to change with position along the axis of a coil? Draw a sketch of B as a function of position, including positions some distance outside the coil, and set it aside.

1. Position two coils so that they are co-axial and touch each other, then set up the Styrofoam blocks and the long ruler as shown in Figure 1. You will perform three sets of measurements of the magnetic field as a function of position along a line parallel to the axes of the coils. The first position is 20 cm to the left edge of the left coil, the last position 20 cm to the right edge of the right coil, for a range of 60 cm. Position the Hall probe (attached to the small ruler) so that it is 20 cm from the left edge of the left coil. Find the position of the probe relative to the large ruler (for example, 12 cm) and write it down.

2. Connect the first coil to the power amplifier output and set up the software to generate a 3-V DC signal.

3. Create a table with two columns, one for the position of the probe and the second for the magnetic field, as described in the software instructions. Enter the starting position of the probe (12 cm, in our example) in the first line of the position column. Since you will perform measurements in 2-cm intervals, enter values in increments of 2 in the following rows of the position column, until you have covered 60 cm (12, 14, 16, ..., 72 in this example).

4. Place the Hall probe at the starting position and tare it. Take your first measurement and record it as described in the software instructions. Move the probe 2 cm along the ruler to the second position in the table, take a measurement, and record. Repeat moving the probe and taking measurements until you have measured the magnetic field at the last position. Then "Stop" the measurement, turn the signal "Off," and rename the "Run" to "B1." Create a graph of the magnetic field as a function of position.

5. Disconnect the first coil and connect the second coil to the power amplifier output so that the current direction in the second coil is the same as it was in the first coil. Add a column to the table for the second set of magnetic field values, and measure the magnetic field for the same probe positions as before. When you are done, label the second set of data "B2." Add the B2 data to the first graph.

6. Connect the two coils in series so that the current flows in the same direction through both. Apply a 6-V DC voltage across the combination and measure the magnetic field for the same probe positions as before. When you are done, label the third set of data "B12." Save and print the figure with all three data sets.

Describe the graph of the strength of the magnetic field B_1 as a function of position and discuss if it is consistent with the magnetic field line pattern in Figure 1.

Describe the relationship between B_1, B_2, and B_{12} and discuss how well your data satisfy the superposition principle of Equation (1).

Question 4   Why was the applied voltage 6 V in Step 6 but only 3 V in Steps 1 and 2?

PART III: ELECTROMAGNETIC INDUCTION

1. Take one of the coils, connect it to the current sensor, and set up the software to measure the current through the coil as a function of time, using a sampling rate of 100 Hz. Please make sure your graph shows the current from the current sensor, **not** the "Output Current" from the power amplifier. Record the current while you move a permanent magnet into and out of the coil. Do this for all possible combinations of speeds and directions (in vs. out, north pole vs. south pole, fast vs. slow).

2. Quantify as best you can how the current induced in the coil depends on the direction and speed of the moving magnet and organize your observations in a table.

3. Place the second coil co-axially with the first so that they just touch each other and connect the second coil to the power amplifier. Take a current measurement at zero applied voltage to establish the "zero current" line for the following measurements. Record the current through the first coil while you apply a voltage signal to the second coil. Use a triangular wave of frequency 1–2 Hz and amplitude 10 V, set the sampling rate to 100 Hz, and use the automatic stop feature to record two periods. Display both the current measured by the current sensor and the voltage output of the power amplifier.

4. Repeat the experiment when the two coils are aligned as before, but some distance apart, and again when they are perpendicular to each other.

5. Quantify as best you can how the current induced in the first coil depends on the distance between the coils and their relative orientation and organize your observations in a table.

Question 5 Since the coils have an ohmic resistance, energy is dissipated when current is flowing. What is supplying this energy in Part III, Step 1?

Question 6 What do you expect to happen to the induced current when you reverse the direction of the current in the second coil, for example by exchanging the connecting wires?

RAY OPTICS, LIGHT INTENSITY, AND POLARIZATION

OBJECTIVES

- To determine the focal length of a lens or spherical concave mirror.
- To investigate the location, orientation, and magnification of images.
- To design a lens combination that produces a desired magnification.
- To measure variations in light intensity.
- To explore properties of polarized light.

INTRODUCTION

The properties of light have been described in terms of rays, waves, and particles. In geometric optics, we think of light as rays that travel along straight lines until they hit an object, like a lens or a mirror, that changes the direction of the rays. The ray model of light is useful for understanding and designing optical instruments. The wave description of light, on the other hand, helps us understand processes like interference and diffraction (see Experiments 10 and 11) and the effects of light polarization. Finally, thinking of light as quantum mechanical particles (photons) is important when investigating details of the interaction of light with

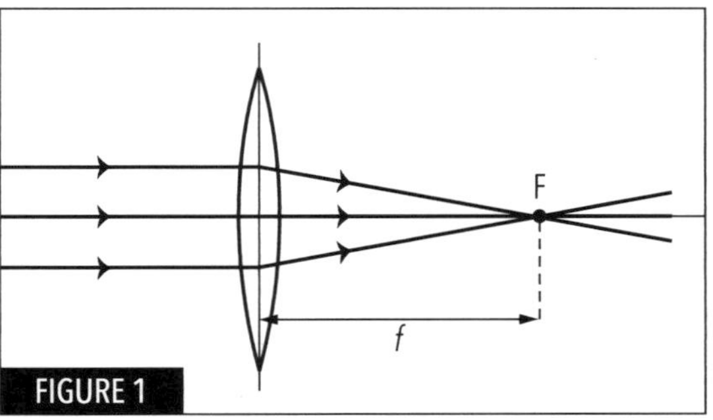

FIGURE 1

Converging lens.

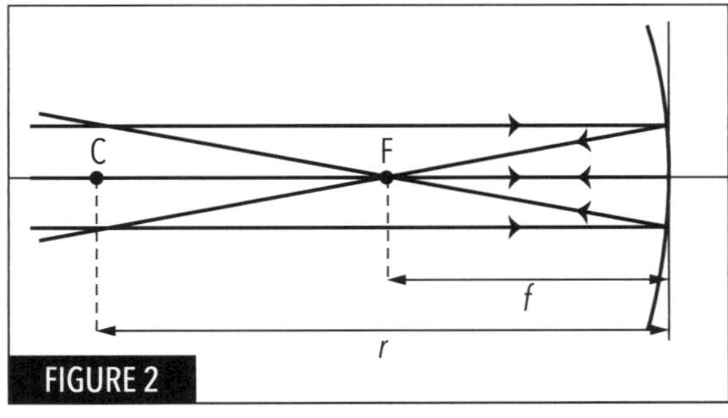

FIGURE 2

Spherical mirror.

materials. In this laboratory, you will perform experiments in geometric optics and explore properties of polarized light by measuring variations in light intensity.

MIRRORS AND LENSES

One characteristic feature of all thin lenses and spherical mirrors is the focal length, f, which is defined as the image distance of an object that is positioned infinitely far way. The focal lengths of a biconvex lens and a concave mirror are shown in Figure 1 and Figure 2, respectively. Notice the incoming light rays from the object are parallel, indicating that the object is very far away. The point C in Figure 2 marks the center of curvature of the mirror. The distance from C to any point on the mirror is the radius of curvature, r. It can be shown that r is twice the focal length.

Figure 3 demonstrates by ray tracing the formation of an image for a converging lens with the object outside the focal length. The object distance d_o and the image distance d_i are the distance between the lens and the object O and the image I, respectively. These distances are related to the focal length f by the lens equation

$$\frac{1}{d_o} + \frac{1}{d_i} = \frac{1}{f}.$$

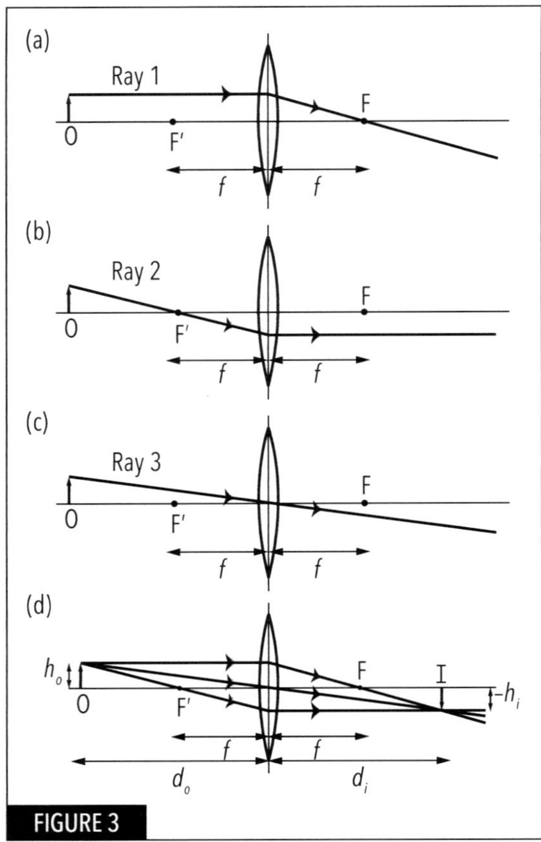

FIGURE 3

Ray tracing for a converging lens with the object outside the focal length. (a) A parallel incoming ray (Ray 1) is refracted so that it passes through the focal point F beyond the lens. (b) A ray incoming from the focal point F' (Ray 2) is refracted so it is parallel to the optical axis beyond the lens. (c) A ray that passes through the center (Ray 3) is unchanged. (d) The point where all three rays intersect beyond the lens shows the image position. The distance between the object O and the center of the lens is the object distance d_o, the distance between the lens and the image I is the image distance d_i. In addition, the object height, h_o, and the negative image height, $-h_i$, are also shown. By convention, the image height is negative when the image is upside down.

(1)

The magnification m is defined as the ratio of image height h_i and object height h_o; m is also related to the ratio of the object and image distances.

$$m = \frac{h_i}{h_o} = -\frac{d_i}{d_o}$$

(2)

By convention, the image height h_i is negative when the image is upside down. The image distance for a lens is positive, when the image is on the side of the lens that is opposite of the incoming light. The focal length of a converging lens is positive.

Figure 4 shows a ray diagram for a concave mirror with the object outside the focal length. The construction of the ray diagram is very similar to the case of the converging lens. The sign convention for the image height (negative for an upside down image) is the same as for the lens, however, the image distance is taken to be positive, when the image is on the same side of the mirror as the object. The focal length of a concave mirror is positive (that of a convex mirror is negative). With these sign conventions, Equations (1) and (2) apply for spherical mirrors.

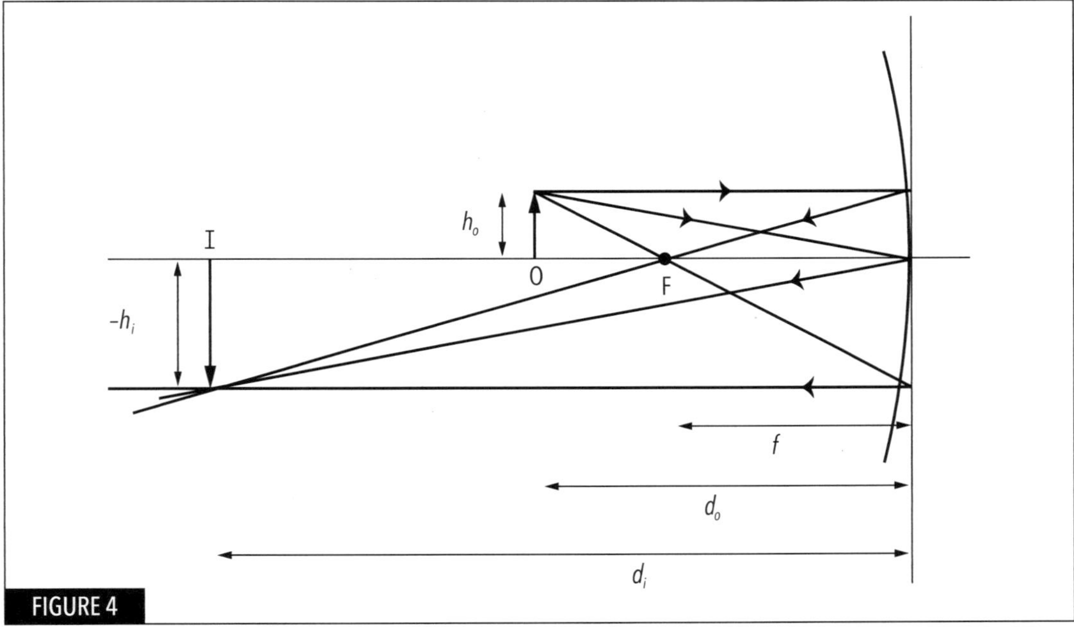

FIGURE 4

Ray tracing for a concave mirror with the object outside the focal length. A parallel incoming ray is reflected so that is passes through the focal point F. A ray incoming from the focal point F is reflected parallel to the optical axis. A ray that reflects off the center is reflected onto itself. The point where all three rays intersect shows the image position. d_o is the object distance, d_i is the image distance, h_o is the object height, and h_i is the image height, which is negative since the image is upside down.

POLARIZATION OF LIGHT

Just like waves traveling on a stretched string, light waves are transverse waves. For a wave traveling on a string aligned with the x-axis, the displacement of a piece of the string from the equilibrium position may be along the y-axis, the z-axis, or any combination of the two. For light waves, the role of displacement from the equilibrium position is played by the electric field. For a light wave traveling along the x-axis, the electric field vector may point in the y-direction, the z-direction, or any combination thereof. The direction of the electric field vector is called the direction of polarization of the light. Light from a source like an incandescent light bulb is unpolarized, this means it is a mixture of light waves polarized in many different directions. Light from computer monitors and hand-held electronic devices, on the other hand, has typically a high degree of polarization. We will use a computer screen as a source for polarized light.

To explore properties of polarized light, we will measure light intensities. Intensity is defined as the average energy passing through a unit area in unit time and has units of watt/m^2. For a round source, such as a light bulb, the intensity decreases with the square of the distance from the source:

$$I(r) = \frac{c}{r^2},$$ (3)

where r is the distance from the center of the bulb and c is a constant with units of watt. If background light is present, it adds a constant to the intensity in Equation (3).

A polarizer sheet is made of a material that allows light of a particular polarization to pass while blocking the other directions. When unpolarized light encounters a polarizer sheet, only the component parallel to the transmission axis passes through and the light behind the sheet is polarized along the transmission axis. When polarized light encounters a polarizer, the amount of light passing through depends on the angle θ between the transmission axis of the polarizer and the polarization direction of the light. The intensity of the light behind the polarizer is described by the equation

$$I(\theta) = I_0 \cos^2 \theta,$$ (4)

where the constant I_0 is the maximum intensity, obtained when the transmission axis and polarization axis are aligned with each other.

Lab Equipment

- Converging lenses
- Concave spherical mirror
- Meter stick, long ruler
- Light source
- Index card

- Holders for lenses, light source, index cards
- PASCO interface and software
- Light sensor
- Polarizer sheets

Part I: Converging Lenses

1. Choose a lens and obtain a first estimate of its focal length in the following way: Choose an object very far away, like a building or a dorm outside. If it is dark outside then use something in the room that is far away. This is done to simulate parallel rays. Parallel rays will be focused to a point at the focal length like in Figure 1. Point the meter stick at the object with the lens mounted between the screen and the object. Adjust the screen to lens distance (d_i) until you see a clear focused image of your object. Record this distance.

Question 1   Use the lens equation with a very large value of the object distance $(d_o \to \infty)$ to show that the image distance recorded in Step 1 is the focal length of the lens.

2. Set up the light source, index card, and lens on the meter stick as shown in Figure 5. Make sure that the distance between the light and the lens is greater than the distance that you measured in Step 1. This makes it possible to get a real, inverted image on a screen (index card) on the opposite side of the lens. Measure and record the object distance (d_o). Then move the screen until you get a clear focused image of the arrow. Measure and record the image distance (d_i). Do this again for two more different object distances for a total of three pairs of data, making sure that at least one object distance is larger and one is smaller than $2f$.

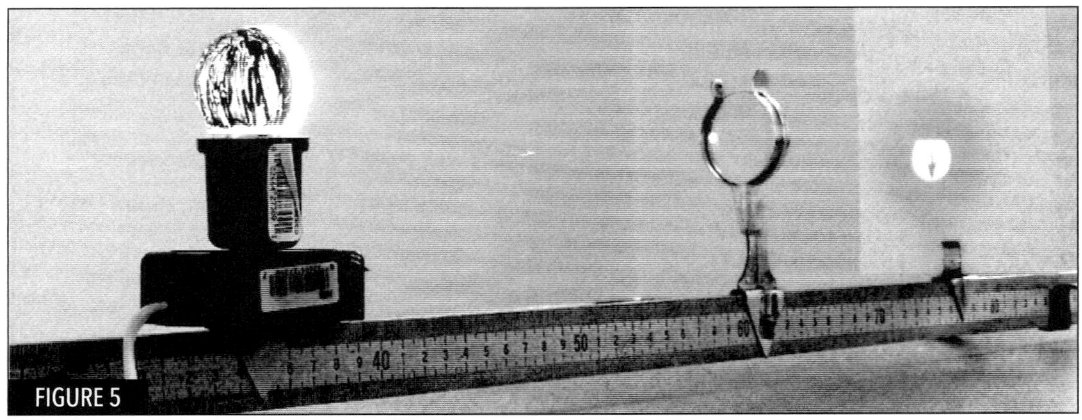

FIGURE 5

Experimental setup to measure object and image distances of a converging lens. An arrow drawn on the other side of the light bulb (not visible) serves as the object; the image of the arrow is visible on the index card beyond the lens.

Question 2   Should the focal length or the magnification depend on how far away the object is placed from the lens? If so, how do they vary with d_o?

3. Enter your data into an Excel worksheet (similar to the chart below) and, using Equations (1) and (2), calculate the focal length (f) and the magnification (m) of the lens for each pair of object and image distances, then calculate the average and standard deviation for f.

    Does this focal length agree with the value obtained in Step 1? If not, what are possible reasons?

	OBJECT DISTANCE d_o	IMAGE DISTANCE d_i	FOCAL LENGTH f	MAGNIFICATION m
Trial 1				
Trial 2				
Trial 3				

Question 3   What happens to the image if you cover half of the lens with an index card? Perform the experiment (be careful to cover the lens, not the object), and record your observation. Explain your observation and draw ray diagrams to support your arguments.

PART II: FOCUSING MIRRORS

1. In this section you conduct similar experiments replacing the lens with a spherical mirror. The first thing to do is find the focal length using a distant object. Place the mirror in a lens holder and point it at an object that is far away. Find the focal length by placing a card in front of and off to the side of the mirror and moving it until you see a sharp image of the distant object. The distance between the mirror and the card is the focal length. Record this length.

2. Set up the equipment as shown in Figure 6. Choose an object distance greater than the focal length you found in Step 1. and determine the image distance by moving the index card until you see a sharp focused image. Measure and record the object distance (d_o) and the image distance (d_i). (Please note, you may have to hold the mirror in your hand and tilt it

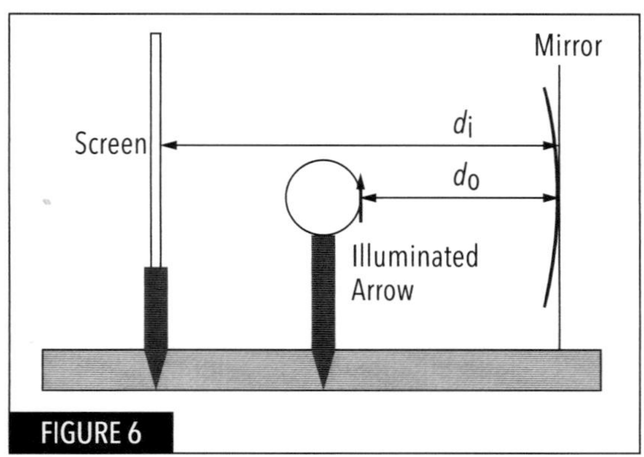

Experimental setup for spherical mirror.

a little to get a focused image of the arrow. If you have difficulty finding the image, calculate the image distance using d_o, f, and Equation (1), then hold the index card at the calculated image distance d_i, and move it a little until the arrow is in focus.) Repeat for two more different object distances.

3. Enter your data into an Excel worksheet and, using Equations (1) and (2), calculate the focal length (f) and the magnification (m) of the mirror for each pair of object and image distances, then calculate the average and standard deviation for f.

Question 4 Is the image on the card a real or virtual image? Is the image inverted or upright?

Question 5 What is the image like when the object is within the focal length? Is it inverted or upright? Enlarged or reduced in size? Hold the spherical mirror in front of your face at a distance smaller than the focal length and record your observations. Is this image real or virtual?

Part III: Lens Combinations

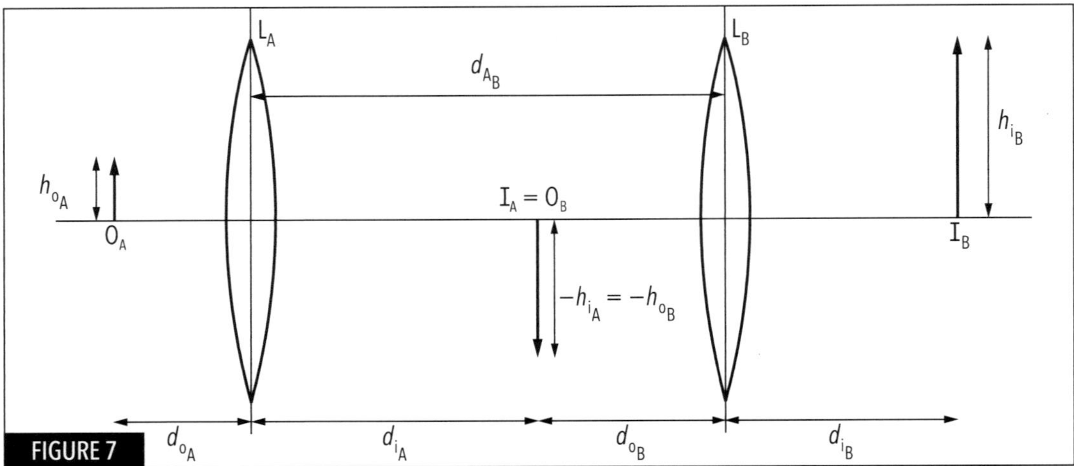

FIGURE 7

Combination of two converging lenses. The object O_A is placed in front of the first lens, L_A. The image produced by the first lens, I_A, serves as the object, O_B, for the second lens, L_B, which produces the final image I_B. Some relevant distances and heights are indicated.

Figure 7 shows a sketch of a setup with the two converging lenses. The total magnification is the ratio of the final image size to the original object height, $m = h_{i_B}/h_{o_A}$, and is the product of the magnification factors of the two lenses:

$$m = m_A m_B. \tag{5}$$

In this section you will design a setup of two converging lenses that creates an upright, real image twice as large as the object.

Question 6   Where are the focal points of the lenses located in Figure 7? Use ray tracing to find the focal points (two each) of both lenses.

1.   Choose a second lens and repeat Steps 2 and 3 of Part I for this lens.

Question 7   Which set of distances, d_{o_A}, d_{i_A}, d_{o_B}, and d_{i_B}, will give a total magnification $m = 2$ for your two lenses? As a first step to find such a set (there are many correct answers, but not all are practical), look at your Excel tables and identify setups with magnifications, whose product is close to 2. Suggest and perform additional experiments if there is no suitable pair of setups in your data.

2. Set up the object and lenses accordingly, locate the first image with an index card to check your first distances, and then find the second image. Measure and record all distances and calculate the total magnification from Equations (2) and (5). Compare your result with the desired value of $m = 2$, refine the target distances with the help of Equations (1), (2), and (5), and repeat experiments and calculations until you have reached $m = 2$.

3. Record your final calculations and target distances d_{o_A}, d_{i_A}, d_{o_B}, and d_{i_B}, compare with the final experimental values, and comment on the agreement.

4. Include a ray diagram of the setup (drawn to scale), calculations, and results in the lab report.

PART IV: LIGHT INTENSITY AS A FUNCTION OF DISTANCE FROM THE SOURCE

1. Connect the light sensor to the interface and set up the software to measure light intensity. Set the gain switch on the light sensor to 1. The light sensor output is in volts, which is proportional to the intensity of the light. The software also gives an option of displaying the intensity in percent. Either choice will work. Set up a table with two columns, the first for distance between the light bulb and the sensor, the second for light intensity. You will be measuring the intensity in increments of 5 cm for distances between 20 cm and 60 cm.

2. Set up the light bulb on the ruler (as in Figure 5) and place the light sensor 60 cm from the bulb so that it points at the light bulb in a direction parallel to the ruler. Take the first measurement, "Keep" it, move the light bulb to the next position, measure, and so on, until you have completed the measurements.

3. Use the software to prepare a graph of the intensity as a function of distance and fit to an inverse square law. Record the equation, parameter values, and uncertainties.

Question 8 In the fit equation of the inverse square law, which parameter corresponds to the background? Which parameter is proportional to the constant c in Equation (3)? What is the significance of the remaining parameter?

Part V: Light Intensity and Polarization

1. Take a polarizer plate, look through it at the light bulb, turning the polarizer plate slowly. Observe any changes in brightness of the bulb. Now take the polarizer plate, look through it at the computer screen, turn the plate slowly, and observe the variation in brightness. The big differences in brightness of the computer screen when looked at through a polarizer are due to the polarization of light emitted by the monitor. In the following experiment, you will measure this intensity variation.

2. Set up a table with two columns, the first for the angle between the transmission axis of the polarizer and the direction of polarization, the second for light intensity. You will be measuring intensity in increments of 30° for angles between 0° and 360°. Keep the gain switch on the light sensor at 1 for this experiment.

3. Resize the software window so that it covers less than a third of the size of the monitor. Open the "ANGLES" document and resize it to fill about two thirds of the screen, being careful not to change the aspect ratio.

4. Note the direction of the transfer axis on the polarizer (indicated by an arrow) and hold the polarizer parallel to the screen so that the arrow is aligned with the vertical (0°) line. Position the light sensor so that it points straight at the polarizer. Take and record your first measurement, then rotate the polarizer until it is aligned with the 30° line, and so on, until you have measured intensities for all angles.

5. Make a graph of your data and inspect it. Compare your experimental results with Equation (4) and discuss the agreement.

6. Use the polarizer plates to examine if other sources in the room emit polarized light. Check light from the light bulb, light reflected off the windows or tables, light from the screen of hand-held electronic devices, and so on. Collect your qualitative results in a table.

10

INTERFERENCE AND DIFFRACTION I

OBJECTIVES

- To investigate wave phenomena with shallow water waves.
- To make direct observations of diffraction and interference.
- To determine the wavelength of a wave from the interference pattern.

INTRODUCTION

Diffraction is the bending of waves around an obstacle. Have you ever wondered why you can hear through an open door when people are speaking in a hallway even though you cannot see them? Sound reaches your ear inside the room since sound waves undergo diffraction at the doorway. The reason for light waves not being diffracted into the room is a mismatch in sizes: The wavelength of sound is comparable to the size of the doorway while the wavelength of light is much smaller than the opening.

Shallow water is an excellent medium to observe wave phenomena. If you drop a pebble into a puddle, you see ripples that show a circular shallow water wave emanating from the point of impact. The ripples that form when you drop two pebbles simultaneously into a puddle show an interference pattern that can be understood with the superposition principle. The waves emanating from the points of impact add up, enhancing each other in some places (constructive interference) and weakening each other in others (destructive interference).

LAB EQUIPMENT

AT EACH LAB TABLE

◎ Ripple tank system consisting of water tank, ripple generator (see Figure 1), and light source

◎ Plane wave actuator (long rod)

◎ Circular wave actuators (bobbers)

◎ Barriers of three sizes

FRONT OF ROOM

◎ Paper towels

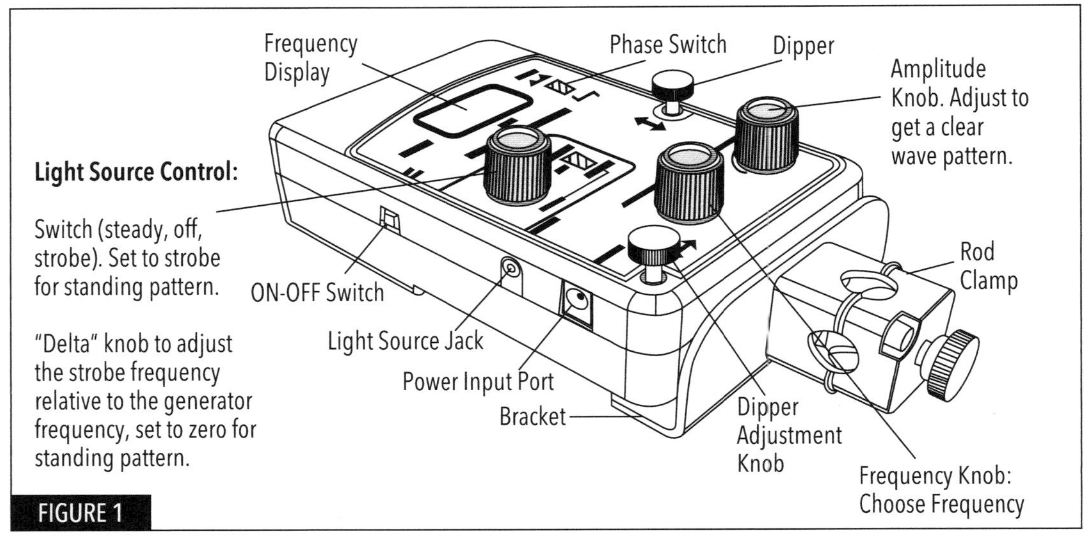

Ripple generator.

The ripple tank contains a thin layer of water in a container with a transparent bottom. The light above the tank serves to illuminate the container, projecting the wave patterns onto paper placed on the table beneath the tank. The ripple generator produces waves of two types; plane waves (with the long rod) and circular waves (with the bobbers). The frequency of the waves is set with the frequency knob and displayed in the LED display. The amplitude of the waves may need to be adjusted to obtain a clear wave pattern. To observe diffraction, barriers are placed directly in the water tank.

Note: *The patterns projected onto the page are "displacement" patterns. In a displacement pattern, bright and dark regions correspond to positive and negative displacements, respectively; gray corresponds to zero displacement. Figure 2 illustrates the difference between a displacement pattern and an intensity pattern for interference between two sets of circular waves.*

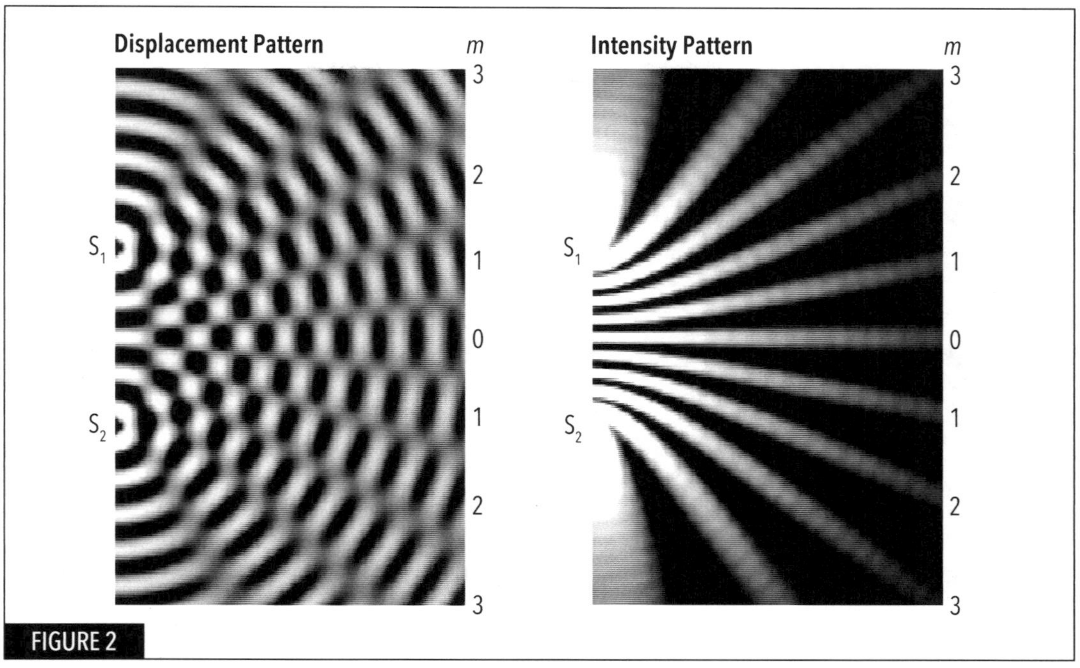

FIGURE 2

Interference pattern from superposition of waves from two point sources S_1 and S_2. The figure on the left shows a snapshot of the total displacement, $D(x,t) = D_1(x,t) + D_2(x,t)$, in gray scale (dark regions correspond to negative displacement $D(x,t) < 0$, bright regions to positive displacement $D(x,t) > 0$, zero displacement is an intermediate gray). This is the displacement pattern. Constructive interference occurs in the stripes with alternating dark and bright regions; m indicates the order of constructive interference. The figure on the right shows the intensity, which is proportional to the time-averaged square of the total displacement and therefore always positive. Bright and dark regions correspond to high and low intensities, respectively. The bright bands indicate regions of constructive interference.

PROCEDURE

PART I: PREPARATION

1. Take a medium sized barrier and measure its length with a ruler. Place the barrier in the water tank and measure the length of the barrier's shadow on the paper underneath the tank. The ratio of the lengths (shadow/actual) is the scale factor s of the projection. Record the lengths and the scale factor in an Excel worksheet.

2. Close the clamp on the water discharge tube. Add about 700 ml of water to the dry ripple tank (please empty the tank at the end of the lab).

Part II: Plane Waves

Generate plane waves with a frequency of 25 Hz. Set the light source switch to "steady" and observe the traveling wave fronts projected onto the paper. Now set the light source switch to "strobe." This will make the wave pattern appear frozen. Keep this setting until Part III, Step 2 of the experiments. Note that the bright and dark areas correspond to the crests and troughs of the waves.

Question 1 Which distances in the pattern correspond to a projected wavelength (bright to bright, bright to dark, ...)?

1. ***Wave Length***

   Record the pattern on paper by marking the center of the dark lines. Measure the projected wavelength λ_p and use the scale factor s to calculate the actual wavelength $\lambda = \lambda_p/s$. Repeat the experiment for two more frequencies and collect all your data and calculations in an Excel worksheet.

2. ***Diffraction from a Semi-Infinite Barrier***

   a. Place a wide barrier in the tank parallel to the agitator rod and in such a way that waves can get past one end but not the other. Generate plane waves with a frequency of 25 Hz and trace the pertinent part of the pattern on paper. Alternatively, you may use the camera on your cell phone to take a photograph of the pattern. If you do, please place a ruler on the table so that it becomes part of the picture and can serve as a scale bar for the image.

   b. Compare the pattern before and after the waves reach the barrier, paying close attention to differences in the pattern in the "shadow" region (behind the barrier) and the open region, and decide if the wavelength changes upon diffraction from the barrier.

   c. Change the frequency to generate waves of a shorter wavelength, trace or photograph the pattern, and compare with the first diffraction pattern.

3. ***Effect of the Size of the Barrier***

   Place a medium sized barrier in the tank and trace or photograph the pertinent part of the pattern. Replace the barrier by a much smaller one; use a narrow pointed object such as pencil tip or a wire. Describe the effect of changing the barrier width on the pattern.

4. *Double Slit*

Place three barriers in a straight line so that they are separated by two narrow slits as shown in Figure 3. Trace or photograph the wave pattern and record your observations for waves of one frequency. You will need to use these data for comparison with those of Part III (circular waves).

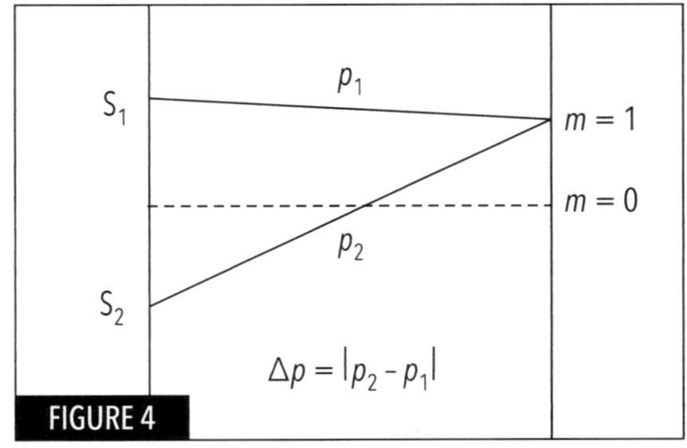

FIGURE 3

Double slit constructed from three barriers. The distance between the centers of the slits is $d = 6.7$ cm, the slit width a is a few millimeters.

PART III: SUPERPOSITION OF CIRCULAR WAVES

1. *Superposition of Two Circular Waves*

Remove all barriers and replace the plane wave actuator by two bobbers to generate two sets of circular waves of the same frequency as in the double slit experiment. Trace or photograph the pattern and compare and contrast the pattern from the double slit with that produced by the two sources of circular waves.

2. *Wavelength from Locations of Constructive Interference*
(set the light source to "steady")

Constructive interference occurs when the path difference (Δp) between two coherent waves is a multiple (m) of the wavelength.

$$\Delta p = m\lambda \qquad (1)$$

Compare your interference pattern from the two sets of circular waves with the pattern in Figure 2 and mark the lines of constructive interference in your pattern. Draw a straight line perpendicular to the central line (maximum of order zero) and identify and mark interference maxima of first,

FIGURE 4

Illustration for path difference determination. Waves from the sources S_1 and S_2 interfere constructively at the location marked $m = 1$. The solid lines labeled p_1 and p_2 are the paths from the sources to the location of the maximum; measure their length with a ruler to find the path difference $\Delta p = |p_2 - p_1|$. The dashed line indicates the central maximum.

second, and third order to the left and to the right of the center along this line. For each of the identified points, note the order of the maximum and determine the distances, p_1 and p_2, to the two sources and the path difference $\Delta p = |p_2 - p_1|$, (see Figure 4).

Question 2 How is the path difference related to the order m of the constructive interference maximum and the wavelength λ of the wave?

For each of the marked maxima, calculate the wavelength from the path difference and the order. Enter your data in an excel sheet and calculate the average and standard deviation of the wavelength. What are possible sources of error?

3. *Effect of Wavelength on Diffraction Pattern*

Question 3 How do you expect the pattern to change with the wavelength of the wave? Predict the location of the first order maxima when the wavelength is reduced by a factor two.

Perform the experiment and compare your measurement with your prediction.

INTERFERENCE AND DIFFRACTION II

OBJECTIVES

- To analyze the diffraction and interference patterns produced by coherent, monochromatic light (laser light) passing through a single slit, a double slit, and around a human hair.

INTRODUCTION

Light from a laser is, to a good approximation, monochromatic and coherent. Monochromatic (of one color) means that the light has a single frequency and wavelength (632.8 nm for the red light of helium-neon lasers); coherent means that light at different points of the beam has a fixed phase relationship.

SINGLE SLIT

When monochromatic, coherent light passes through a narrow slit, it is diffracted and spreads into the "shadow" region behind the barriers. (You saw this effect with water waves passing between two barriers in the previous experiment, *Interference and Diffraction I.*) For the single-slit experiment with light waves, the size of the screen and the distance between the slit and the screen are very large compared to the slit width. This allows us to observe an interference pattern due to interference between light rays passing through different parts of the slit. Figure 1 shows a schematic of the single-slit experiment and a part of the intensity pattern observed on the screen.

To understand the first minimum, $m = 1$, consider one ray in the upper half of the slit and a second ray in the lower half of the slit, a distance $a/2$ from the first, as shown in Figure 2. When these rays are diffracted by an angle θ, the path difference between them is $\Delta p = p_2 - p_1 = (a/2)\sin\theta$. If they are exactly out of phase, the path difference between them is half a wavelength, $\Delta p = \lambda/2$, and the two rays interfere destructively. Equating the two expressions for Δp, we find $\sin\theta = \lambda/a$.

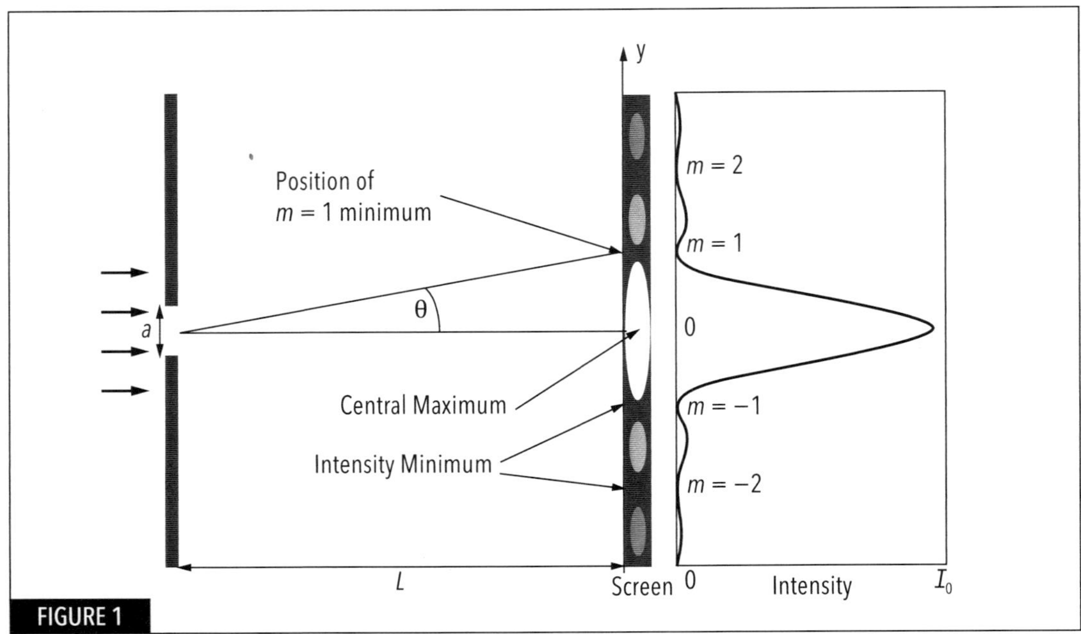

FIGURE 1

Illustration of the single slit experiment, not to scale. Parallel rays (plane waves) of monochromatic light fall on a slit that is perpendicular to the rays and has width a. The light is diffracted by the slit and illuminates a screen at a distance L behind the slit. The pattern on the screen shows bright spots where the intensity is high, separated by dark regions near the intensity minima. The graph on the right shows the intensity as a function of position on the screen; I_0 is the maximum intensity and corresponds to angle θ = 0. The intensity minima are labeled by integers $m = \pm 1, \pm 2$; note that the m = 0 minimum is missing. The angle θ is related to the position y on the screen and the distance L between the slit and the screen through tanθ = y/L.

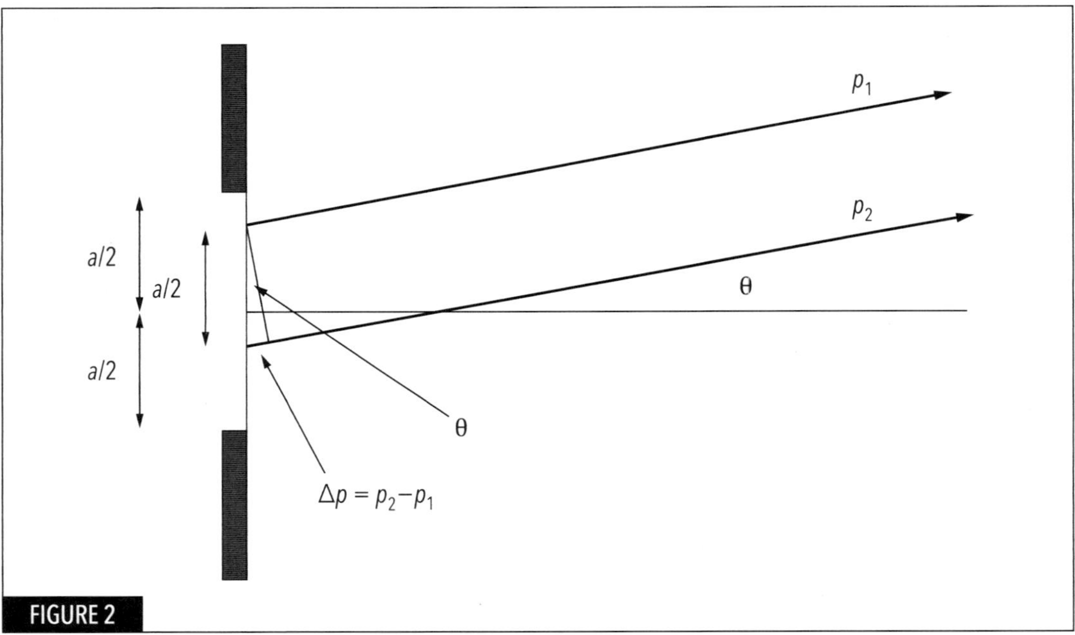

FIGURE 2

Illustration for the path difference between a pair of rays leaving the slit from positions in the slit a distance a/2 apart. The small right triangle indicated near the slit shows that $\Delta p = (a/2)\sin\theta$, which is true for small angles θ.

Since the path difference is the same for all pairs of rays that are separated by $a/2$ in the slit (you should convince yourself of this) all the pairs interfere destructively leaving the screen dark for this angle. Similarly, to understand the $m = 2$ minimum, divide the slit into four quarters and consider rays leaving the slit a distance $a/4$ apart; this yields $\sin\theta = 2\lambda/a$ for the angle. In general, we find minima for angles θ with

$$\sin\theta = m\,\frac{\lambda}{a}. \quad m = \pm 1, \pm 2, \dots \qquad \text{Condition for Single Slit Minima} \quad (1)$$

Double Slit

When a double slit is brought into the path of a laser beam, two effects contribute to the interference pattern observed on the screen. The first effect is interference between the light rays passing through the individual slits. (This is similar to the interference you observed in the water tank with two point sources.) The second effect is diffraction due to the finite width of each slit. Figure 3 illustrates the path difference for waves coming from two point sources (infinitely narrow slits) when the screen is very far away.

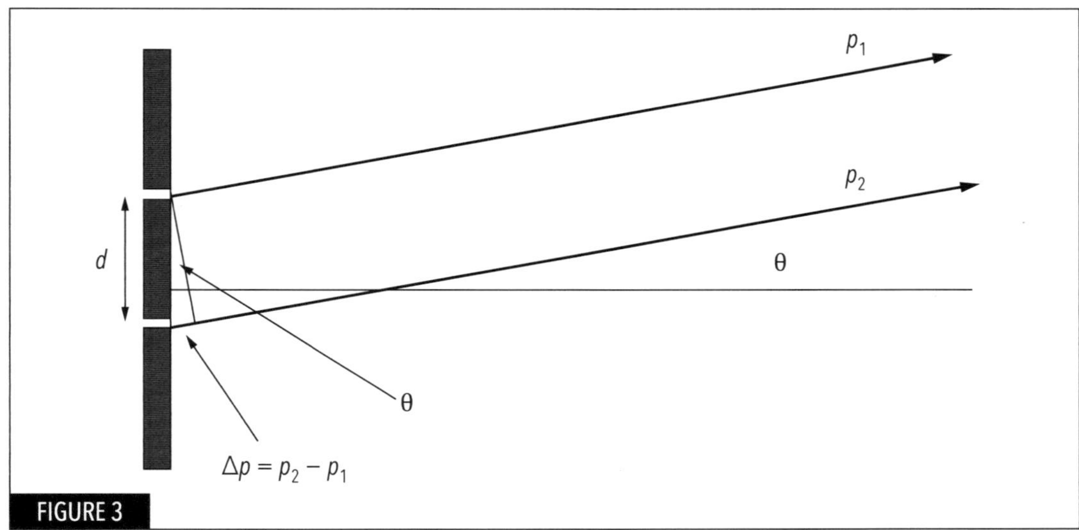

Illustration for the path difference between rays passing through narrow slits a distance d apart. The small right triangle shows that $\Delta p = d\sin\theta$, which is true for small angles θ.

Destructive interference between the rays from the two narrow slits occurs when the path difference is an odd multiple of half a wavelength $\Delta p = (2n + 1)\,\lambda/2$. Since the path difference is related to the angle θ through $\Delta p = d\sin\theta$, we find minima for angles θ with

$$\sin\theta = \left(n + \frac{1}{2}\right)\frac{\lambda}{d}. \quad n = 0, 1, 2, \dots \qquad \text{Condition for Double Slit Minima} \quad (2)$$

Since each of the slits has a finite width a, interference also occurs between rays from an individual slit. Those interference minima occur at angles given by Equation (1). Figure 4 shows a schematic of the double-slit experiment and a part of the intensity pattern observed on the screen.

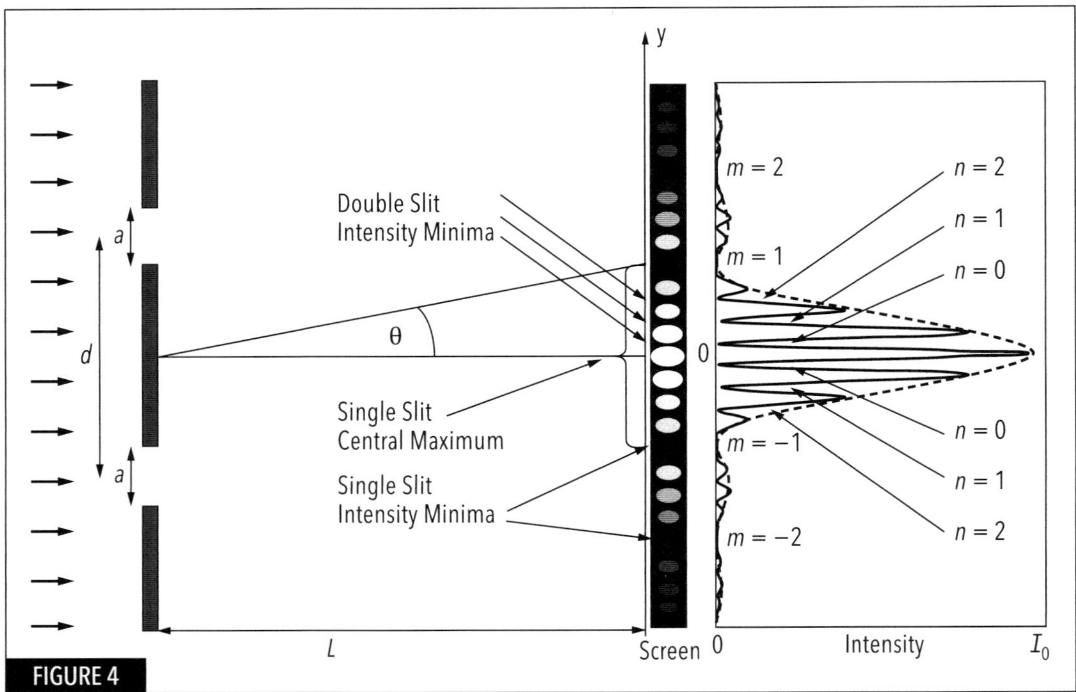

FIGURE 4

Illustration of the double slit experiment, not to scale. Parallel rays (plane waves) of monochromatic light fall on two slits of width a separated by a distance d. The light is diffracted and illuminates a screen at a distance L behind the slits. The pattern on the screen shows bright spots where the intensity is high, separated by dark regions near the intensity minima. The solid line in the graph on the right shows the intensity as a function of position on the screen; the broken line shows the intensity that would be observed if there were only a single slit of width a; I_0 is the maximum intensity and corresponds to the angle $\theta = 0$. Intensity minima due to the double slit interference are labeled by integers $n = 0, 1, 2$, minima due to the finite slit width are labeled by integers $m = \pm 1, \pm 2$. The angle θ is related to the position y on the screen and the distance L between the slit and the screen through $\tan\theta = y/L$.

The distance L between the slits and the screen is much larger than the distance between minima on the screen. (When you do the experiment, use a distance of at least 1.2 m.) Therefore, we may use a small angle approximation $\sin\theta \cong \tan\theta$ when evaluating the experiments.

Lab Equipment

◎ Laser

◎ Magnetic optical bench

◎ Slide holder

◎ Slides with various slit configurations

◎ Screen

Procedure

Part I: Single Slit Experiment

Question 1 Starting from Equation (1) show that the positions y_s of the single slit minima are given by

$$\frac{y_s}{L} = m\frac{\lambda}{a}, \quad m = \pm 1, \pm 2, \dots \quad (3)$$

when the small angle approximation $\sin\theta \cong \tan\theta$ is used.

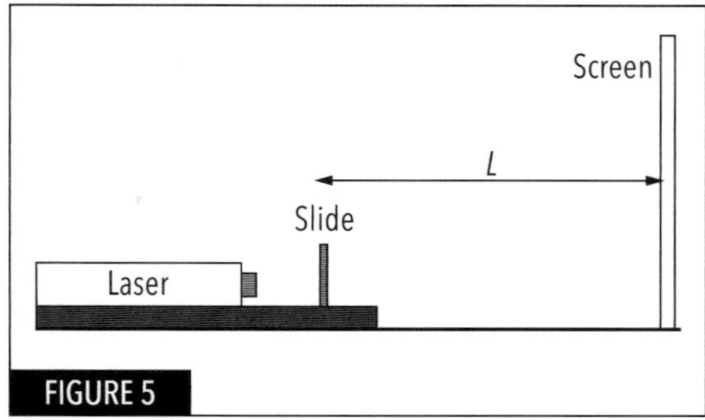

Experimental setup for single and double slit experiments. The distance L between the slide and the screen should be at least 120 cm. Please make sure the laser beam is perpendicular to the slide.

1. Insert a slide with a single slit of width a = 0.08 mm and align the laser, the slide holder and the screen so that you observe a diffraction pattern with minima of about 20 orders on the screen. Copy the pattern to a piece of white paper and measure the distance between the slide and the screen. Measure the distance $2y_s$ between the minima of order 10, calculate the wavelength from Equation (3), and estimate the uncertainty in the wavelength from the uncertainty in your measurements.

Question 2 Predict the position on the screen of the m = 5 minimum if you replace the single slit of width a with a slit of width $a/2$.

2. Perform the single slit experiment with a slit of width a = 0.04 mm and compare your result with your prediction.

Part II: Double Slit Experiment

Question 3 Starting from Equation (2) show that the positions y_d of the double slit minima are given by

$$\frac{y_d}{L} = \left(n + \frac{1}{2}\right)\frac{\lambda}{d}, \quad n = 0, 1, 2, \dots \quad (4)$$

when the small angle approximation $\sin\theta \cong \tan\theta$ is used.

Question 4 Derive equations for the slit width and the slit separation for the double slit experiment.

1. Insert a slide with a double slit and align the laser, the slide, and the screen. Copy the pattern on the screen to a piece of white paper and measure the distance between the slide and the screen. Compare your intensity pattern with the pattern shown in Figure 4 and label the minima due to double slit interference and single slit diffraction. Determine the width of the central single-slit maximum and count the number of fringes (bright spots) inside this maximum. Calculate the slit width and slit separation and compare with the values printed on the slide.

Question 5   Predict how the slit width affects (a) the width of the central single-slit maximum (b) the number of fringes in the maximum.

Question 6   Predict how the slit separation affects (a) the width of the single-slit central maximum (b) the number of fringes in the maximum.

2. Repeat the experiment for three more double slits and make a table of your results. For each of the four double slits, enter the slit width, slit separation, width of the central single-slit maximum, and the number of fringes in the central maximum. Compare your results with your answer to Question 5 and Question 6.

Part III: Measuring the Thickness of a Hair

Take a piece of hair, tape it to a slide holder, and align the laser and the slide holder so that you see a diffraction pattern on the screen. The diffraction pattern should be similar to a single-slit diffraction pattern; this is an example of Babinet's principle, which is explained in your textbook. Determine the thickness of the hair from the diffraction pattern and estimate the uncertainty in your results.

Question 7   Does hair of different color differ in thickness?